MW00465626

ROCKS FOSSILS AND DINOSAURS

WWW.ICONPUBLISHINGGROUP.COM

For information write to:
Icon Publishing Group, LLC
Noble, Oklahoma 73068
www.iconpublishinggroup.com

First Edition, First Impression. Printed in Canada.

All Scriptural passages used in this book that are not New King James Version are properly annotated.

The photograph used for the cover was provided by Triebolt Paleontology in Woodland Park, CO and was used with permission.

Library of Congress Cataloging-in-Publication Data:
Sharp, G. Thomas
Rocks, Fossils and Dinosaurs/G. Thomas Sharp

ISBN: 978-1-60725-587-1

For information about other creation books by this author, please visit our website: **www.creationtruth.com**

Dedicated to Dr. Jim D. Standridge,
A dear friend, a man of Biblical
and Spiritual insight, a genuine
prophet of God.

Table of Contents

Foreword

Immediately upon receiving Christ as my Savior, I was given a New Testament which I devoured in four days. I then began reading through the entire Bible. What an eye opener! Like so many of my generation, I was thoroughly indoctrinated by the evolutionary propaganda that was fashionable in the mid to late sixties. I cannot even begin to describe how morally and spiritually devastated I was when I first "bought the lie." I simply connected the dots: If evolution is true, everything can be explained and accounted for without reference to God. If God is not needed to explain anything, then God is not needed. If God is not needed, I had no reason to even believe that God existed, much less to trust in Him or live my life in a way that would be approved by Him.

In my senior year of high school, I had a class called Physical Anthropology, which turned out to be nothing more than a kind of catechism for human evolution. After "seeing the pictures" of our supposed "less than human" ancestors (including the artists rendering of their social life and interactions), I took a "required to graduate" sociology class, which was no more than a continuation of the indoctrination I was receiving in my earlier class on evolutionary anthropology. My next class was psychology, where

I "learned" that because there was no need for God, there was no need for the "invented moral standards." I was, therefore, to consider myself just another animal and free to do as I pleased. I was not the only one of my generation "getting the message," though not everyone took evolutionary dogma to its logical conclusion. I did, and I was carried along in the current of a world seemingly without God and without hope.

Then, through a simple but clear witness about our Savior and what He did on the Cross and accomplished in His resurrection, I came to believe that Jesus Christ was who He claimed to be and did what He claimed to do. I turned to Him in faith and my life was radically and dramatically changed for the better and forever. But what about all the so-called "science" I had learned that seemed to exclude God from the "real world?" Fortunately, one of the men that God used to witness to me was Pastor Chuck Smith of Calvary Chapel of Costa Mesa, and he provided me with my first New Testament as a believer and gave me a copy of a wonderful little book by Dr. Henry Morris titled The Bible and Modern Science. As I was learning the wonderful truth as it is found in God's Word, I also began learning about the monstrosity of a lie that is called "macro" or Darwinian evolution.

Now, with forty plus years of the Christian life and ministry behind me, I can say without a doubt that God's Word is true in all that it affirms. I discover this to be so even on matters of scientific importance, and the corresponding truth that is called "macro evolution" is a very big lie. These are incredibly illuminating and liberating truths. If there was ever a day in which the Christian community needs to be armed with these inseparable truths, it is now. If there was ever a generation that needed to know the truth about ultimate origins, it is this generation. Parents should not assume that if they send their kids to Sunday school—a good thing to do—they will be protected when they send them to College. Even some "Christian" colleges and universities have bought into the evolutionary lie or some version of the Christian faith that tries to accommodate it. There is no need for our young people to be sent as lambs for the academic slaughter. If they read what is found in Rocks, Fossils and Dinosaurs by Dr. Thomas Sharp, they will have all the ammunition they need to stand for the truth and stand up to the lie.

You do not need to be a scientist or to specialize in one of the sciences to understand the facts and the implications of those facts with regard to ultimate origins. But, if you are not a scientist, you will probably need a little help if you desire to understand what we are all up against. Dr. Sharp, in very readable and easy to understand terms, breaks it down for the layman. I have been reading (with great interest) the writings of both Creationists and Evolutionists for more than four decades. I can assure you that I am not easily impressed. What I found in Rocks, Fossils and Dinosaurs did impress me. This is not a book just for the college and university student who is likely to be challenged (in a very direct way) regarding Creation and Evolution. This is a book for anyone and everyone who wants to understand what is at stake. Dr. Sharp has already proved to be one of the most able defenders of the Christian faith in the face of the evolutionary challenge. Rocks, Fossils and Dinosaurs should prove to be a "Creationist Classic." If you are headed for college or university, "Don't leave home without it."

—George Bryson, Director
Calvary Chapel Church Planting Mission

Overview

One is forced to conclude that many scientists and technologies pay lip-service to *Darwinian Theory* only because it supposedly excludes a Creator from yet another area of material phenomena, and not because it has been paradigmatic in establishing the canons of research in life sciences and the earth sciences.

Michael Walker, "To Have Evolved or Not To Have Evolved: That Is The Question?" *Quadrant*, October 1981, p. 45.

Introduction

The dinosaur story, as told today in the classical evolutionary context, is among the top two or three conditioning tools effectively used to dissuade Christian youth against Biblical faith. This is particularly true for those between the ages of three and ten years. Throughout the western world, billions of public and private dollars are spent annually to both sustain the "dumbing-down" conclusions of this story, and to create new ways to tell it!

There is a continuous conforming factor impacting believing parents and educators surrounding the present dinosaur story. This is due to a terrible academic hook imbedded in the story that alleges some scientific credibility for the story. This fact has been completely ignored or misunderstood for years by the Bible believing community. **Ignored**—because Christians have been beguiled to believe the advertised notion that state school curriculum was credible, accurate, and scientific. **Misunderstood**—because we have not taken the time to know what the Bible really says about these kinds of issues, nor do we fully appreciate the limitations of empirical investigation (modern science).

The dinosaur story, as told today in the classical evolutionary context, is among the top two or three conditioning tools effectively used to dissuade Christian youth against Biblical faith. This is particularly true for those between the ages of three and ten years. Throughout the western world, billions of public and private dollars are spent annually to both sustain the "dumbing-down" conclusions of this story, and to create new ways to tell it!

There is a continuous conforming factor impacting believing parents and educators surrounding the present dinosaur story. This is due to a terrible academic hook imbedded in the story that alleges some scientific credibility for the story. This fact has been completely ignored or misunderstood for years by the Bible believing community. **Ignored**—because Christians have been beguiled to believe the advertised notion that state school curriculum was credible, accurate, and scientific. **Misunderstood**—because we have not taken the time to know what the Bible really says about these kinds of issues, nor do we fully appreciate the limitations of empirical investigation (modern science).

This condition is best described by the prophet Hosea who warned that, God's "people are destroyed for the **lack of knowledge...**" (Hosea 4:6). It is interesting to note that the phrase "*lack of knowledge*" is most simply translated "*because of their ignorance.*" It is of further interest that the word "*destroyed*" (Hebrew, *damah*) means to be **dumb** or **made silent**. Thus, Hosea perfectly captured the nature of this modern problem. He said that God's people are silenced or speechless; and that they are, in effect, destroyed in this matter of dinosaurs and other origins issues because they are ignorant of salient truths that completely exonerate the Biblical view. If they only knew the limitations of genuine scientific processes and what the Bible actually has to say on the matter, most, if not all, of the conflict would immediately disappear.

I have said for years that if we really knew what science can and cannot do and if we knew what the Bible really says on origins topics, evolution would not be a problem for true believers.

The Apostle Peter tells us to:

> Sanctify the Lord God in our hearts; and be ready always to give an answer to every man that asks you a reason for the hope that is in you...(I Peter 3:4-5, KJV)

There is no greater application of Peter's command than for Christian parents and educators to declare inwardly and outwardly that the Lord Jesus and His Word is truth. To declare with Francis Schaeffer that Biblical reality is total truth, "not just a series of truths but Truth—about all of reality" (*Christian Manifesto*, pp. 19-20).

This declaration must be followed by appropriate and disciplined learning processes necessary to make ourselves, our children, and our students "*ready*" to give answers to those who ask for reasons about our hope. I emphasize that the greatest application and need for the effective use of this spiritual formula is with our own sons and daughters. This is imperative because what we believe about our origin dictates what we believe about God, our purpose, our identity and our destiny. This is Biblical worldview!

There was a time in our Judeo-Christian past when general Bible information was all that was needed for a proper Biblical defense, but that day is over! In fact, it has been over for a long time. The consistent and thorough decline of the Bible throughout the twentieth century continues at a greater pace today and has ultimately destroyed all vestiges of America's Judeo-Christian ethic given us by our Puritan Forefathers. As a result, today's American societal mores and values are significantly pagan and anti-Biblical.

Because this is true, it ups the ante for the Bible believer. In order for a twenty-first century disciple to be sufficiently "*ready*" to give an answer, as Peter admonished, he or she now needs additional preparation in academic areas in which we are generally uncomfortable. Today, to be an effective witness/evangelist, especially to the pagan/secular population of our culture, requires that we become knowledgeable in Genesis 1-12 so that we can effectively defend this foundational truth.

Thus, questions and their appropriate answers about dinosaurs, for example, become continuously more important for the defense of the twenty-first century believer.

For the most part, the Biblical view of dinosaurs is one of the great secrets in today's world because the secular view is all that is being presented to us and to our children. This mass indoctrination has reached such a consummate volume in our culture that the very appearance of a dinosaur skeleton militates against the existence of the Creator God and the Biblical account of creation.

Over the past twenty years, I have had hundreds of parents, teachers, and pastors tell me they simply do not know how to deal with questions about dinosaurs and the Bible (and this is equally true for most origins issues). They have confessed that when asked they essentially have nothing to say and pass it off with statements like, "I am not really sure what to say about these creatures, however, they probably lived a long time ago, maybe in the 'millions of years' that existed between Genesis 1:1 and Genesis 1:2." Some Pastors glibly say dinosaurs are not important to the gospel, so whenever they existed God alone knows. Thus, we see the tragic application of Hosea's prophecy being fulfilled right in front of our very eyes.

The dinosaur story as told by the secular community carries with it a ton of faith-killing baggage, all of which contends with the historical accuracy of Genesis and the legitimacy of the death, burial, and resurrection of Jesus. We have painfully **missed** this implication for years, primarily because most believing parents and teachers were schooled themselves in evolutionary dogma. Furthermore, because we have not taken the time to think through the obvious conclusions of the evolutionary story for ourselves to make sense out of this dilemma, many well-meaning Christians have attempted some kind of Biblical concession, or some kind of humanistic interface, or some kind of attempted reconciliation with the evolutionary story.

Charles Darwin

I've repeatedly said that God did not send Charles Darwin to Earth to explain the book of Genesis to the Church—the Holy Spirit will do that if we will let Him. Nevertheless, somehow, Darwin has been allowed to make significant inroads into our thinking, all of which is contrary to the Bible.

This little book is written in a non-technical style (though all the conclusions contained herein are based on the best technical understanding available today). The driving motive for having written this book is I Peter 3:15 —that readers of this book will discover answers to questions that may have puzzled them for a long time.

It is my desire that readers who are non-scientists—particularly students, parents and pastors—will be able to profit from the message contained in these pages. All terminology and technological jargon will be properly defined in the text, and a complete Index has been prepared for the reader's convenience.

I have intentionally covered more than dinosaurs because arguments in the origin's debate are driven by polarized belief systems. With little exception, the reader will be able to understand how to apply Biblical argument to most all origins issues.

This book is not meant to be an exhaustive study of the subject. And because it isn't, I have carefully documented all the cited authors, who are the professionally trained experts in their various fields, so the reader will be able to obtain their printed matter and use it as a guide for further study. Readers are encouraged to contact Creation Truth Foundation to secure any of this material. It is my prayer that every reader will be strengthened for having read this book.

I must acknowledge those who helped with the production of this book, without which it could not have been accomplished. The CTF Staff, particularly Bob Dugas, and the wonderful assistance of Blue Channel

Media were priceless, especially Gary Dressler and Thomas A. Sharp, my son. I must not fail to mention the marvelous grammatical assistance I received from Bonnie Jackson. I am indebted to many of the dear saints of Immanuel Baptist Church of Skiatook, Oklahoma, for their prayers and support, especially the angel of that House, Dr. Jim D. Standridge. The continuous prayer and the many words of encouragement from my Pastor, Dr. Haskell D. Rycroft were irreplaceable. I am deeply indebted to the faithful support of Mrs. Sabrina Miller. I must give special thanks to those who reviewed this book and gave timely suggestions and recommendations: Colonel John Eidsmoe, Dr. John Morris, Mr. J.R. Hall, Dr. Gary Parker, and Dr. Charles Jackson. Above all, the faithful support, care, and patience of my dear wife of 46 years, Mrs. Diane F. Sharp. Thanks to you all!

—G. Thomas Sharp

Dr. Henry M. Morris is the twentieth century father of the revival of Biblical Creation, upon whose shoulders we all stand.

1

Science Falsely So-Called

The limit of man's imagination is unending in a world that is guided by evolutionary relativism. This is especially true in the present mysterious domain of dinosaur hunters. Nothing punctuates this reality more vividly than the recent "scientific discovery" (or more appropriately "evolutionary" blunder) of a fossil that was alleged to be a dinosaur with feathers.

Now before you break out in sidesplitting laughter, let me fill you in on the amazing details—it gets funnier. It seems a team of 007-type smugglers from China, driven by profit rather than science, both easily and inadvertently hoodwinked the already overwrought imagination of several prominent evolutionary scientists with a concocted fossil. This prestigious group of scientists already believed birds came from dinosaurs so they were easy to fool.

A principle in this group, Mr. Stephen Czerkas of the Dinosaur Museum in Blanding, Utah, purchased the fossil in question from a private dealer in February 1999. The Sidney Morning Herald reported Czerkas' reaction upon finding the fossil:

> "It was stunning," Mr. Czerkas recalls. "I could see right away that it didn't belong on sale. It belonged in a museum." He hastily contacted a patron who put up the $80,000 the dealer was asking... and took his prize home in a state of high excitement, **convinced he had discovered evidence of a pivotal moment in evolution** (emphasis added).

On October 15, 1999, Czerkas and Professor Xing Xu of the Institute of Vertebrate Paleontology and Paleoanthropology (Beijing, China), jointly appeared at a news conference in Washington, D.C., in support of the fossil (Austin, 2000, March).

Then without legitimate peer review and in the highest evangelistic fervor of modern evolution, *National Geographic* (November 1999) published the entire story—as only they can—under the fantastic title: *"Feathers for T. Rex?"* The exquisite pictorial that began the article was a two-page, full-color spread that included the caption: "It's a link between terrestrial dinosaurs and birds that could actually fly" (Sloan, 1999). The author of the article, C. P. Sloan, indicated: "...we can now say that birds are theropods just as confidently as we can say that humans are mammals. Everything from box lunches to museum exhibits will change to reflect this revelation" (Sloan, 1999, p. 102).

Obviously, they so wanted to believe that they had finally found a "missing link"—one that firmly proves their far-fetched belief that modern flying birds actually came from some reptilian ancestor—that they didn't take the time to corroborate the validity of their strange fossil. They named the fossil *Archaeoraptor* (meaning ancient or primitive bird of prey).

But the plot thickens as the power of man's unsanctified imagination reveals itself once again. This is what happened. Prior to the November 1999 *National Geographic* article, at the October 1999 meeting of the Society of Vertebrate Paleontology, Dr. Storrs Olson, said, in effect, to publish opinions about *Archaeoraptor* without sufficient peer review is disingenuous.

In fact, Dr. Olson wrote an open letter to the officials of *National Geographic* on November 1, 1999, warning them that they were headed for worldwide embarrassment if they printed their article about the fossil (Storrs Olsen, *"Open Letter,"* November 1, 1999, to Peter Raven at *National Geographic* Society; http://www.trueorigin.org/birdevoletter.asp). Raven is the Senior Scientist at NGS. Dr. Olson is the Senior Scientist and Curator of the Bird Division of the National Museum of Natural History in the Smithsonian Institution, Washington DC.

However, the reader must balance Dr. Olson's rejection of the interpretation of this evidence fossil, even though he is a convinced evolutionist himself, with the fact that many evolutionists believed the interpretation to be true. *Science News* noted in their January 15, 2000, issue that the fossil was a major find for evolution even though many scientists had concerns about the fossil appearing to be manipulated—the body of a bird and the tail of a lizard (which it turned out to be). Nevertheless, the *National Geographic* authorities somehow convinced themselves that the two parts of the fossil belonged together.

To make matters much worse, officials for *National Geographic* have since acknowledged that they received Dr. Olson's letter of warning but chose to ignore it because they believed Olson's comments were based on rumors. Thus, another "Piltdown fairy tale" is given scientific sanction long enough to fool millions of unsuspecting people (The Piltdown man fiasco was the apparent discovery of a human-like skull and a few teeth in Piltdown, England, told to the public for forty years to be an authentic "missing link," which turned out to be a deliberate fraud).

Olson plainly stated in his open letter to Peter Raven:

> The idea of feathered dinosaurs and the theropod origin of birds is being actively promoted by a cadre of zealous scientists acting in concert with editors at *Nature* and *National Geographic* who

> themselves have become outspoken and highly biased proselytizers of the faith. Truth and careful scientific weighing of evidence have been among the first casualties in their program, which is now fast becoming one of the grander scientific hoaxes of our age—the paleontological equivalent of cold fusion. (Olson, 1999)

This is the tragedy associated with these apparently premeditated blunders of evolutionary conclusions, and there have been many of them. Once a ton of feathers has been released in a hurricane, their recovery is impossible. Think about it this way, between October 15, 1999, and January 21, 2000, *National Geographic* exhibited this contrived fossil in their Explorer's Hall in Washington, D.C., and 110,000 people (the majority of them children) saw the fossil and heard the story. Most, if not all, of these children have never heard the truth about this fossil (and never will). They will go to their death believing they actually witnessed a "missing link." Consequently, they will live under the faith-killing influence of another evolutionary hoax.

2

Death: It's Origin and Purpose

Why is the evolutionary story of dinosaurs so harmful for believing sons and daughters? The answer to this question is incredibly serious and may surprise you. Its harm is due to the negating impact it has on the Biblically stated reason for which Christ died!

But, you ask, "How can that be true?"

Let's think about this for a minute. The evolutionary explanation of dinosaurs goes something like this. About 200 million years ago, some kind of ancient amphibian provided the evolutionary pathway, they believe, that led to the development of a unique group of carnivorous reptiles that began to rule Earth (they are said to be unique from other reptiles because of the special morphology or form of their pelvic girdle). We are told that it was this unique evolution that brought on the age and rule of dinosaurs.

Many of them are said to have possessed such a savage, physically dominant manner until they are presented to the public as ruthless predators able to attack and consume all other animal life. (Of course, some of these special creatures are known to be herbivores which, evolutionists contend, provided most of the food for their carnivorous cousins.)

The time of their appearance on Earth, we are told, is restricted to the Mesozoic Era. The evolutionary time frame for this Era is believed to be 230 to 65 million years ago. Of course, this chronology is totally determined by the evolutionary interpretation of the rock record.

Why did they appear? Evolutionary theorists don't have the foggiest idea why they appeared on Earth because they have never found a common ancestor for which an airtight case can be made. However, we must keep in mind that the entire story is based on their faith in the reality that evolution actually happened and their interpretation of the rock record. These two facts are important for you to remember.

The story continues by alleging that this menacing supremacy of these "terrible lizards" (the actual meaning of the word "dinosaur") continued for about 135 million years, or until about 65 million years ago. At this point, they assert that some horrible event or circumstance suddenly, and finally, ended the age of dinosaurs and their rule on Earth. Some evolutionary theorists believe the rock record shows several mass extinction periods throughout the geological ages, and this reveals that dinosaurs were diminishing from Earth several million years before the supposed final catastrophic event that closed the Cretaceous Era. All evolutionists, however, believe dinosaurs to be extinct.

They continue their hypothetical story by maintaining that once the dinosaurs disappeared from the earth, it was another 60 to 64 million years (depending on which evolutionist you are reading) before the first man came walking out of the jungle to take his place in the great evolutionary drama and history of life. It is maintained that during the age of the dinosaurs' **death, pain and suffering** was commonplace on Earth millions of years before the first man arrived on the scene.

THE ORIGIN, CAUSE AND PURPOSE OF DEATH

Here's the dilemma! The evolutionary view of dinosaurs causes an unmanageable Biblical contradiction, not only with the Genesis accounting of creation, but with the Biblical Gospel as well. Specifically, it causes an irreconcilable problem with the Biblical origin of death, the reality and origin of sin, and the purpose for the cross of Jesus Christ. The Apostle Paul makes it quite clear in Romans 5:12 that death was a result of the first Adam's sin for "...just as through one man sin entered the world, and **death** through sin...."

He tells us again in I Corinthians 15:21, that "...by man [the first Adam] came **death,** [and] by Man [Jesus Christ, the last Adam] came the resurrection of the dead."

In other words, Paul obviously taught that the Biblical view of **death** is both crucial to and directly connected with Christ's first coming, His death, His burial, and His resurrection. Therefore, Adam's sin and subsequent death is integrally associated with the Biblical gospel.

The Bible emphatically tells us that both physical death and spiritual death are due to Adam's sin. The evolutionary theorists tell us that physical death is a developmental mechanism leading to higher and more complex living systems.

If, indeed, dinosaurs were on Earth killing air-breathing, blood-containing life before man arrived on Earth, and if this caused millions of years of pain and suffering, then both Paul and Genesis are wrong! This evolutionary assumption impacts the historical accuracy of Genesis because Genesis 1 tells us that all animals were originally created to eat plants (Genesis 1:29). Of greater consequence, this view destroys the historical purpose and the theological basis for Christ's death, burial, and resurrection (the gospel). If Adam's sin did not bring death, the Biblically stated reason for Christ's crucifixion is simply not true. Moreover, the origin of sin and evil become seriously obscured.

What do you think Paul meant in I Corinthians 15:20-22:

> But now Christ is risen from the dead, and has become the firstfruits of those who have fallen asleep. For since by man came death, by Man also came the resurrection of the dead. For in Adam all die, even so in Christ shall all be made alive.

He obviously meant that Adam's sin was the cause of death and that before Adam sinned there was no death. But, wait a minute, what kind of death is being suggested here? Is it both physical and spiritual death, or is it just spiritual death? Does this death include day-five and day-six animals, or is it exclusive to mankind? Were there death, pain, and suffering before Adam sinned?

Now, I have heard arguments that allege Adam's sin only caused spiritual death, not physical death, and that only Adam and Eve were affected by this curse, not animals. These arguments are proffered in an attempt to explain the possibility that the physical death of animals and other humans existed before Adam sinned during a time called the pre-Adamic Earth. This would allow for an interminable time period over the course of the creation week in an attempt to justify millions of years of development and Earth history in Genesis 1.

However, Paul makes it very plain in the above passage that the "first" Adam caused both physical and spiritual death. And, furthermore, that through Christ (the last Adam), we now have victory over the sting of death (physical death) and the hope of eternal life (which is restored spiritual life that Adam lost in the Garden of Eden). Remember, the curse in Genesis 3 included both physical death and spiritual death (see Genesis 3:9-20).

Moreover, it seems unambiguously clear to me that Genesis 1 says the Creator saw the creation to be "good," even "very good." This means that everything created—including the heaven, the earth, and the sea—plus all of their contents—were perfect and that divinely so.

Death of any dimension (pain, suffering, fatigue, disease, etc.) in this environmental context is inconceivable!

I know that a quibble is offered about the digestion of plants being a form of death. But plants were specially designed to provide food for both man and animals and were uniquely created to do so (Genesis 1:20-30). Plants do not possess the "breath of life," in Hebrew called *nephesh* life (compare Genesis 1:21, 24, 28; 2:7&19). In every instance of their creation, both man and animals are called *chay nephesh*, which when translated from its Hebrew means "living creature, living soul, breath of life," etc. The point is, plants do not possess this special level of creation; and, as such, they do not possess the same biological make up and purpose as do animals and man.

All *nephesh* life needed oxygen and had flowing blood to carry oxygen throughout their bodies. Remember, Moses tells us that "the life of the flesh is in the blood" (Leviticus 17:11). It is a clear Biblical fact that there was no **death** of *nephesh* life before Adam sinned. Dr. James Strong explains:

> This idea is especially clear in the creation account, in which God tells man that he will surely die if he eats of the forbidden fruit (Genesis 2:17—the first occurrence of the verb). **Apparently, there was no death before this time.** When the serpent questioned Eve, she associated disobedience with death (Genesis 3:3). The serpent used God's words, but negated them (Genesis 3:4). **When Adam and Eve ate of the fruit, both spiritual and physical death came upon Adam and Eve and their descendants (cf. Roman 5:12).** They experienced spiritual death immediately, resulting in their shame and their attempt to cover their nakedness (Genesis 3:7). Sin and/or the presence of spiritual death required a covering, but man's provision was inadequate; so God made a perfect covering in the form of a promised redeemer (Genesis 3:15) and a typological covering of animals skins. (Strong, p. 152; emphasis added)

Furthermore, the writers of the *Theological Workbook of the Old Testament* add:

> The normative OT teaching about death is presented in Genesis 3:3 [which is a reference to Genesis 2:17], where God warns Adam and Eve **that death is the result of rebellion against His commands.** Since God's purpose for our first parents was never-ending life, the introduction of death was an undesirable but necessary result of disobedience. The physical corruption of the human body and the consequent suffering and pain brought about by the Fall were only the obvious symptoms of death. **Death is the consequence and the punishment of sin. It originated with sin.** (Harris et al., 1, p. 1169; emphasis added)

But what happened to the animals? One fact is for sure, man's sin required animals to die. Where do you suppose the skins came from if not from animals? I think it is very consistent with the divine intent of the creation, and for the optimal provision of those whom God created in His own image and likeness (man), to provide them with every means for their comfort and support. Therefore, it seems only natural that if man fell, all aspects of the creation fell with him, including the animals. There can be no other meaning to the Apostle Paul's statement in Romans 8:19-22 where he tells us that the "**whole creation** groans and labors with birth pangs until now," because it was subjected to vanity (futility) and has been delivered into the "bondage of corruption."

There should be no question about the fact that Adam's disobedience caused God to curse the whole creation along with Adam and Eve. Moreover, the fact that God chose the "skins" of animals with which "to cover" them properly (or to provide atonement for them), also indicates that Adam's sin brought death to the animals. Because Adam sinned, the whole creation changed, and among the realities of this change was the introduction of mortality of men and animals (Genesis 3:14-19).

There is one last fact supported by Scripture that is pertinent to this discussion—the origin of carnivory. There could not have been carnivorous behavior before Adam's sin, because both animals and man were originally provided herbs, seeds, fruits, and grasses for their food (Genesis 1:29-30). Carnivorous behavior was only possible after the curse. But how long was it after the curse? It must have been by the time of the Flood, or just after it. Nonetheless, "death" entered the creation because of Adam's sin, and it was not the result of "meat-eating" caused by dinosaurs, as some contend.

Sadly, many evolutionists understand the debate over the origin and purpose of death better than most professing Bible believers understand it. G. Richard Bozarth, an atheist, wrote in the *American Atheist* all the way back in February of 1978 that:

> Christianity has fought, still fights, and will fight science to the desperate end over evolution [he used the term "science" in this sentence rather loosely, he obviously meant naturalism], because evolution destroys utterly and finally the very reason Jesus' earthly life was supposedly made necessary. **Destroy Adam and Eve and the original sin, and in the rubble you will find the sorry remains of the Son of God.** If Jesus was not the redeemer who died for our sins, **and this is what evolution means,** then Christianity is nothing! (p. 30; emphasis added)

NOTE TO THE READER: *When the evolutionary crowd refers to "science" in context with origins issues, they are generally referring to "naturalism" and calling it science, as did Bozarth in this quote.*

Simply stated then, Paul is right about the origin, cause, and purpose of death or the secular scientific community is right—they both cannot be right. Why did Jesus die on the cross and rise the third day? He did so to reconcile believers to the Father, but He also destroyed the sting of death at the same time (I Corinthians 15:55-57). The first Adam sinned and death was the result (both spiritual and physical death). If this is not true, the cross-work of Christ means absolutely nothing.

The last Adam came to reconnect fallen man to his Creator, to atone for man's sins, to provide covering for man's nakedness, and to bring man access to God, which was the terrible price vicariously laid upon our sinless Savior (Genesis 3:15). Accordingly, if dinosaurs were alive millions of years before the first man appeared on the earth and they were freely hunting and killing air-breathing, blood-containing *"nephesh"* life, then the gospel is a cruel joke! So, ideas do have consequences—some of them have devastating consequences to the legitimacy of the gospel of Jesus Christ!

3

Trusting the Bible's History

In this little volume, we will pursue our research in the manner suggested in John 3:12. Here Jesus asked Nicodemus an exceedingly important question:

> If I have told you **earthly things** and you do not believe, how will you believe if I tell you heavenly things? (emphasis added)

In the first place, Jesus is telling Nicodemus that our faith is objective. Bible-believing Christians, by divine design, were not intended to be some marginal group of uninformed, subjective, non-foundational, primitive, religious neophytes. So they must not act as if they are! Biblical Christianity, with all its doctrinal corollaries, is not some pie-in-the-sky superstition or medieval mythology that was founded by some cave-dwelling mystic whose name was lost in antiquity. Biblical Christianity is expressly objective; it is established on clear, demonstrable facts having a unique and brilliant Founder whose character is flawless and whose Word we can perfectly trust.

Jesus is provoking Nicodemus concerning the objective nature of Christianity. You will notice that Jesus admonishes Nicodemus to "check" those things He has said about Earth. In essence, Jesus is saying that if the earthly record, particularly the Genesis record, is not checkable and verifiable, then we really have no temporal basis upon which to trust His promises about heaven, either. (It is important for the reader to note that evolutionism is the number one reason today for the rejection of the second advent of Jesus Christ, a fact that I will explore a little later.)

One of the exercises in which we will engage in this book is a careful checking of Earth in order to compare the predictions of the Bible—especially regarding rocks, fossils, and dinosaurs—with what we actually observe around us. This is critically important because if Jesus is saying anything in this verse, He is telling us that what we think about Earth and its history provides evidential believability for our thoughts about our destiny.

Jesus said if you do not believe the things I've said about Earth—things you **can** check—then how will you believe the things I've said about heaven—things you **cannot** check! This gives significant meaning to the words of the Apostle Peter (I Peter 3:15) when he tells us that the ability to effectively answer questions that we are asked also supports the integrity of our hope in Christ. Therefore, our failure to answer the questions does two things. It strengthens the skeptic's position, and it weakens our own hope. Is it any wonder that so many professing believers are confused, miserable, and hopeless?

It is, therefore, fascinating for me to see in John 3:12 that Christ is suggesting that we can objectively consider what He has said about Earth. Thus, He helps us understand in this verse that our faith can have an objective base in these matters. It's not just a blind leap in the dark. It is not based in the whims of some misguided guru. And because of this Scriptural fact, we at Creation Truth Foundation use this verse for the basis of our ministry's motto:

Teaching disciples to trust the Bible's history for its accuracy, so they will trust the Bible's promise for their destiny.

Throughout this book, therefore, we will check the Scriptural predictions of Earth as they relate to rocks, fossils, and dinosaurs. If these predictions check out, or if they agree with observational reality, then we have a significant, objective basis to trust the accuracy of the Bible's history. But more than that, we also have another reliable hope that the Bible's promise about Heaven is true as well.

We have been so overdosed with evolutionism and "wowed" by the pseudo-academic claims of secular science in its support, that most of us are convinced subjects like history, biology, geology, paleontology, astronomy, and anthropology, to name just a few, can only be accurately taught in some secular, state school by secular, state-trained teachers. If this is true, then it means that the church and the Bible are only profitable for religious issues, specifically, matters of faith and salvation and not matters of history and science.

It means that the Bible's history and science are not dependable accounts, that they must be ignored in order to maintain an informed and scientific status in today's world. However, when I hear these secular assertions, I merely respond that if the conclusions of your "educated" point of view lead invariably to the godless evolution of matter and man, taking billions of years, and having no purpose or design, this point of view requires more faith than the Biblical view.

To allege that Earth is nothing special—that it and the people on it are really lost in space without purpose or destiny, and that Earth is just a speck of meaningless dust without a redemptive purpose—is, in my judgment, a classic case of godless materialism. This is a strategy of the War of the Ages, having neither academic foundation nor scholastic proof. It is not education at all, it is a tortured belief system purporting itself to be academic, but whose only effective support is the mere force of biased political censorship and rabid indoctrination.

4

Earth History and Enmity

Most people do not realize that the evolutionary community has constructed a naturalistic view of dinosaurs (as well as the many other evolutionary ramifications) from their interpretation of rocks and fossils, as I have already said. Specifically, the so-called geological ages developed from the assumptions of uniformitarianism, which is a natural prediction of Darwinian dogma. Conversely, the Biblical creation community has developed their view of the dinosaurs, etc. based essentially on the content of Genesis 1-12 along with other supporting Scriptural passages. In short, the entire origins debate consists of beliefs held by faithful proponents—they may be beliefs about dinosaurs or any other origins subject—the debate is not about science at all. It is a battle of interpretations based on two diametrically opposing worldviews.

Therefore, the first cause for the life and the extinction of dinosaurs, cannot possibly have the same scientific status as, for example, the study of genetics, gravity, or protein synthesis because it simply does not have the same scientific rigor or measurability available to research.

The dinosaur question is in the special world where interpretation is everything, and the crowd who has the most media influence gets their story told most.

Both the creationist and the evolutionist study the very same data but from different preconceptions, thus producing completely different interpretations. The secular world interprets dinosaur data from the naturalistic model while the creationist proponent interprets the same data from the Biblical Creation and Flood Model. It's really a war of worldviews! It is not a battle between the Bible and science (never has been); and because it is not, one of the greatest deceptions developed in the twentieth century that is still with us today is the notion that the creation/evolution debate is between the wild-eyed, uninformed, fundamentalist fanatics and the highly educated, erudite specialists of secular, naturalistic science.

There is absolutely no scientific sanction to either view. Both models are belief systems that are held strongly by their respective proponents, but neither is capable of scientific proof. Oxford Zoologist L. Harrison Matthews in, of all places, the Introduction to a 1972 edition to Darwin's book, *The Origin of Species,* made this fact unmistakable:

> In accepting evolution as a fact, how many biologists have paused to reflect that science is built upon theories that have been proven by experiment to be correct, **or remember that the theory of animal evolution has never been thus proven?**...The fact of evolution is the backbone of biology...biology is thus in the peculiar position of being a science founded on an unproved theory—is it then a science or a faith? Belief in the theory of evolution is thus exactly parallel to belief in special creation—both are concepts which believers know to be true but neither, up to the present, has been capable of proof. (Matthews, 1972; emphasis added)

To better illustrate this fact, not long ago I was asked to participate in a discussion that aired on the PBS Network in Oklahoma City concerning the fairness and feasibility of the Tulsa Zoo displaying a creationist exhibit on their premises. The panel of which I was a part consisted of a highly qualified company of university professors (all evolutionists) from the University of Oklahoma, Oklahoma State University and Oklahoma City University, plus a cynical moderator. I rather felt like Daniel in the lion's den. Immediately at the outset of the program—cameras rolling—the professor from Oklahoma University made the declaration that all of the evidence favored the evolutionary view. Because anyone can make this allegation, I simply asked him for an example of some of this evidence. He responded by providing me a litany of evolutionary assumptions, again without any corroborative evidence. He said that evolution is plainly supported by comparative anatomy, biochemistry, the fossil record, and genetics, etc. I quickly told him that while all of this was interesting, could he please specifically cite some of that evidence from any of these fields to which he was referring. He then said that "everyone" knows that chimpanzees are the nearest relative to humans in the Animal Kingdom. Of course, he believed, as did all of the evolutionary participants, that man was a member of the Animal Kingdom. Now, where did that assumption originate? Because, he continued, it has been proven that their biochemistry is about 99.4% similar to humans.

Finally, we were getting to the real point of the discussion. I told him I essentially agreed with his data, except for the fact that some recent scientific discussions are indicating the possibility of lesser percentages of similarity between humans and chimps, but that I explicitly disagreed with his interpretation of this evidence. This is always the rub—the interpretation of the evidence.

The truth is to a greater or lesser degree, humans are biochemically similar to many biological creatures. Nevertheless, I told him (and the audience) that this same biochemical data could also mean that a masterful Designer, as the Creator God identified in Genesis, used the same basic design for all living organisms. And the fact that chimps and man have a similar body style and live on the same planet with similar needs, it's only reasonable to expect this Master Engineer to make their biochemistry similar. Therefore, this evidence could also mean that man and chimps have a common Designer or a common Creator, and not a common ancestry at all. The subject abruptly changed!

I pray the reader sees in this anecdotal experience the difference between true scientific fact and one's interpretation of the fact that is always colored by preconception (this is a practice that is always associated with the origins debate for both creationists and evolutionists). Quite simply, in origins discussions, we are most often exposed to a biased interpretation that is presented as fact. This is why I often say, particularly in this context; there is a great difference between the opinion of a scientist (his faith, bias, or worldview) and scientific opinion (the results of bona fide research). In this connection, I wonder how many of us have stopped to reflect who it is that produces textbooks for all state schools.

So, in view of this circumstance, it will be important for us to give some attention to the history and the development of the modern discipline of Earth Science, especially regarding how and why the features of Earth are interpreted the way they are interpreted. When you look at the surface of the earth, how do you explain what you see? How do you explain, for example, why fossils occur, or how old a rock or fossil is, or how (limestone) solution caves and canyons formed, or, for that matter, what the rock record actually means.

But, more importantly, **why** do you explain Earth the way you do? Here, as in all origin's topics, we are always exposed to one's interpretation. What we see is always based on what we believe about what we see, and the evolutionary scientist is not exempt from this practice.

It is highly interesting and intriguing, I think, that evolutionary paleoanthropologist (a person who studies human fossils), Roger Lewin, wrote in his 1987 book *Bones of Contention* that:

> Preconceptions are rarely acknowledged because this, after all, would be "unscientific." And yet preconceptions are an individual scientist's guide to how to view the world with a degree of order that allows structured questions to be asked. The anonymous aphorism [adage] **"I wouldn't have seen it if I hadn't believed it" is a continuing truth in science.** (Lewin, Simon and Schuster, 1987, p.19; emphasis added)

Here's the point: Our preconceptions are totally based in what we believe to be true about a matter, and this is extraordinarily true in origins issues. The truth is no one observed the origin of the universe, the origin of life, or the origin of man. So, how does anyone propose to scientifically explain these phenomena? Origins issues, from a pure scientific standpoint, are mysteriously wrapped in a riddle and all tied up in a conundrum. We simply do not know from a pure scientific point of view how and when life came into existence! What we say about these areas of discussion is simply what we believe about them. It is as Professor Lewin said, *"I wouldn't have seen it if I hadn't believed it."* This means that all of us have presuppositions about origins from which we interpret the evidence.

Moreover, what we believe and accept as truth (naturalistic evolution or Biblical creation) forms the basis of our interpretation and reveals our ultimate worldview. While someone's discussion may be dressed up in academic and scientific clothes, nowhere is preconception or presupposition more evident than in the origin's discussion, whether creationist or evolutionist. For example, there is very little justification for the advertised geological column that appears in most secular Earth Science textbooks. Nowhere is this fact exhibited any better than in an article written by geologist John Woodmorappe:

> Eighty to eighty-five percent of the Earth's land surface **does not have even 3 geologic periods appearing in "correct," consecutive order**...a significant percentage of every geologic period's rocks does not overlie rocks of the next older geologic period...Since only a small

> percentage of the Earth's surface obeys even a significant portion of the geologic column, **it becomes an overall exercise of gargantuan special pleading and imagination for the evolutionary-uniformitarian paradigm to maintain that there ever were geologic periods.** The claim of their having taken place to form a continuum of rock/life/time of ten biochronologic "onion skins" over the Earth is therefore a fantastic and imaginative contrivance. (Woodmorappe, pp. 46-71; emphasis added)

Professor Woodmorappe is simply telling us that the so-called uniformitarian rock column is nothing more than a contrivance of the evolutionary marketers (but more on this later). We must remember, the Apostle Peter told us in II Peter 3:3 that "scoffers" would arise in the last days who would "willfully ignore" evidence for the Genesis Flood.

To ignore the palpable evidence for this global, catastrophic event of the recent past is to leave yourself without an understandable mechanism for explaining the many features associated with the surface of the earth (its rock layers, fossils, canyons, caves, mountains, etc.); that is, unless you add the evolutionary baggage of hundreds of millions of years of Earth history. (Please understand that this is not to say that the residual catastrophism since the Flood could have also added to the accumulation of Earth's present features. It probably did.) Nevertheless, this is what the secular scientific worldview has done and it is from this "godless" view of Earth that evolutionists also establish their explanation of dinosaurs—millions of years and all! And it is the millions and billions of years that the evolutionist cannot give up.

In this regard, it is interesting that Charles Darwin penned in a letter to J. Croll in January of 1869: "I am greatly troubled at the short duration of the world according to Sir William Thompson (Lord Kelvin) for **I require for my theoretical views a very long period before the Cambrian formation** (Darwin, F., 1972, p. 163; emphasis added).

Creation: Mind-Set of Early Scientists

Georges Cuvier

Most all scientists from the seventeenth century to the early nineteenth century were creationists, as well as believers in the realities of the global Flood of Genesis 7-8. However, there were a few scientists such as the French anatomist Georges Cuvier who believed there were a series of local or regional floods, rather than a single worldwide Flood event. This view is similar to the view of secular scientists today who are more and more abandoning all aspects of the doctrine of uniformitarianism.

Please note that the term **uniformitarianism** refers to a naturalistic model in modern geology that was popularized in the middle to late nineteenth century and hardly finds support today. Pro-evolutionary influences then asserted that the features of the earth are developed in a gradual manner using only natural forces of erosion, chemical weathering, freezing and thawing, among other natural explanations as agents of topographical change. Thus, requiring millions of years to accomplish the many features we now observe on the surface of the earth. However, most of the modern scientific disciplines were founded by researchers who were also Bible believing creationists. Dr. Henry Morris makes this fact very clear:

> The basic compatibility of science with Christian theism is even more obvious when it is realized that modern science actually grew in large measure out of the seeds of Christian theism. It is absurd to claim, as modern evolutionists often do, that one cannot be a true scientist if he believes in creation...most of the great founders of science believed in creation and, indeed, in all the great doctrines of Biblical Christianity...Men such as Johann Kepler, Isaac Newton, Robert Boyle, David Brewster, John Dalton, Michael Faraday, Blaise Pascal, Joseph le Clerk Maxwell, Louis Pasteur, William Thompson

(Lord Kelvin), and a host of others of comparable stature were men **who firmly believed in special creation and the personal omnipotent God of creation, as well as believing in the Bible as the inspired Word of God and in Jesus Christ as Lord and Savior.** Their great contributions in science were made in implicit confidence that they were merely "thinking God's thoughts after Him," and that they were doing His will and glorifying His name in doing so. (Morris, 1984, p. 31; emphasis added)

Johann Kepler

The Age and Meaning of the Rocks

Isaac Newton

During the Dinosaurs, Design, and Destiny worldview seminars that Creation Truth Foundation conducts each year in many churches and schools in the United States, I am repeatedly asked about the age and meaning of rocks. These questions arise because most church members, like most people of the twentieth and twenty-first centuries, have been taught that the rock record supports millions of years of Earth history as well as a "simple to complex" evolution of life.

Most do not have access to an unrevised account of the history of science covering the period 1700-1900, and, as a result, their understanding is limited to the information found in secular textbooks. Therefore, they think these evolutionary notions of geology are the objective consensus of the best science available on the subject today.

Thus, the evolutionary assumptions about the earth are considered unquestionable by most Christians, or they're not sure what to think.

Some want to believe the Bible account of history, but they are not sure they can trust the Bible's history because of the influence of secular science to the contrary. Thus, as a practicing Christian, they struggle with a literal reading of the Genesis record. Of course, the state sponsored presentation of evolution also permeates the content presented by all other informational outlets in the American culture, including state and federally funded cultural centers, radio, television, newspapers, other media sources, and the Hollywood entertainment industry to name just a few. But is the evolutionary information presented by these organizations truly scientific as claimed?

I recently received an email from a critic of Biblical creation who chided me for not knowing what he called "the history of the rock record." He said that by now everyone should know that the vast ages of the rock layers are uncontested because they have all been accurately dated by radiometric processes.

I returned a brief rebuttal of his assumptions and told him that almost everyone knows, except maybe him, that the present geological time scale was arbitrarily established, assigned ages and all. It is very much the same as it was in the early to mid 1800s (it just presents older ages), and, of course, this was a long time before radioactivity was discovered. Really, everybody should know this fact.

Edmund M. Spieker wrote, in the *Bulletin of the American Association of Petroleum Geologists:*

> ...I wonder how many of us realize that the time scale was frozen in essentially its present form by 1840. How much world geology was known in 1840? A bit of Western Europe, none too well, and a lesser fringe of North America, **all of Asia, Africa, South America, and most of America were virtually unknown.** How dared

> the pioneers [of this theory] assume that their scale would fit the rocks in these vast areas, by far most of the world? Only in dogmatic assumption—a mere extension of the kind of reasoning developed by Werner from facts in his little district of Saxony. And in many parts of the world, notably India and South America, **it does not fit (the geologic time scale). But even there it is applied!** The followers of the founding fathers (of geology) went forth across the earth and in Procrustean fashion (insensitive of the possibility that differences may exist) made it fit the sections they found, even in places where the actual evidence literally proclaimed denial. **So flexible and accommodating are the facts of geology.** (Spieker, 1956, p. 1803; emphasis added)

What does all of this mean? From what source did the idea of millions and billions of years of earth history come? How do evolutionists justify the "goo to you" evolution of life? Is there anything observable or testable about any of these notions?

I will attempt to answer these and other questions in the balance of this book. But know that the entire debate is simply another example of **The War of the Ages** that was predicted in Genesis 3:14-19 but most do not understand the historical significance of that idea either. So, let me briefly explain.

THE BATTLE OF ENMITY

At the time of the great Curse in Genesis 3, God pronounced a continuous, seething **enmity** or hatred (Genesis 3:15) that would be fought between the "seed of the woman" and the "seed of the serpent" until the end of time. This has caused perennial warfare between good and evil. This War can also be described as conflict between truth and error, or light versus darkness, or order against disorder, or creation and evolution, or Kingdom versus Kingdom, etc. Moreover, this War has a number of nuances. Therefore, I am convinced that any idea or notion that denies the inspirational origin or the absolute inerrant accuracy of God's Word is without foundation. This would include the Bible's accompanying doctrines and practices, especially the fact that Christ is Creator and Messiah. Anything or anyone that attempts to deceive or weaken the redeemed of God (God's elect) regarding the above two Biblical realities, must be seen as an aspect of the inherent evil associated with the "seed of the serpent" whose agenda is the propagation of the "mystery of iniquity." Moreover, it is due to these two overarching belief systems and their respective leaders (the Lord Jesus Christ and that serpent of old, called the devil and Satan) that the War of the Ages is fought, and will be fought until the end of time. No human being is exempt from this War! You are either actively engaged or you are already defeated; there are no gray or neutral areas.

Jesus made this fact very clear when He said, *"He that is not with Me is against Me, and he that gathers not with Me scatters abroad"* (Matthew 12:30). In every generation there have been, and will continue to be, many different battles of this War; all fought on different fronts, with what seem to be different causes or purposes. But careful analysis will reveal the true motive or cause in the conflict to be just another aspect in the long War between good and evil. The reader must understand that the debate between creation and evolution is a battle in this Great War. There are many aspects of a Biblical worldview and creationism is just one of them. However, it is the foundational tenet of all Biblical faith and, as such, it is the most important. This is why it is so necessary to be able to give an answer concerning the Biblical view of Earth and why it is of equal importance to recognize and refute the evolutionary interpretation promoted by the "seed of the serpent" regarding the history of the earth.

THE HISTORY OF UNIFORMITARIANISM

I don't want to beat a dead horse, but for the sake of those who may read this book and don't fully understand the transitions in geological thought over the last 200 years, this section is vital. This is especially

James Hutton

true as regards modern evolutionary geology. I want to discuss the vagaries inherent in this very important historical record because I think they hold the key to understanding the ultimate confusion that exists in the present naturalistic opinion of Earth's surface. I want to add further emphasis to the intellectual poverty associated with "deliberately ignoring" evidence for the Flood.

In 1785, James Hutton (1726-1797), an agriculturist with some medical training, first published his *Theory of Earth* which questioned the legitimacy of the Mosaic chronology of earth. Hutton did not agree with the Genesis account that dates the age of Earth in thousands of years; he believed the rocks had to be much older than this. To accept this conclusion, he also denied the historicity of the Genesis Flood. Hutton's views, when popularized by Sir Charles Lyell, ultimately led to a wholesale departure of the Biblical view of Earth beginning in the nineteenth century and throughout the twentieth century.

Charles Lyell

This process involved many people and it took only thirty to forty years to dominate many in European science. Evolutionary geologist Carl O. Dunbar reports his view of the historical development of this model:

> The uprooting of such fantastic beliefs [that is, belief in the Genesis Flood] began with the Scottish geologist, James Hutton, whose **Theory of the Earth,** published in 1785, maintained that **the present is the key to the past,** and that, given sufficient time, processes now at work could account for all the geologic features

> of the Globe. This philosophy, which came to be known as the **doctrine of uniformitarianism** demands an immensity of time; it has now gained universal acceptance among **intelligent and informed people.** (Dunbar, 1960, p. 18; emphasis added)

Please understand, contrary to Dunbar's commentary, Hutton was not a professional geologist. In fact, most of the personalities that figured into the establishment of this "new" geological doctrine had professions outside the field of geology. Dr. Henry Morris highlights this fact:

> ...the basic structure of modern historical geology was worked out over a hundred years ago by such men as **James Hutton (an agriculturist with medical training),** John Playfair (a mathematician), William Smith (a surveyor), **Charles Lyell (a lawyer),** Georges Cuvier (a comparative anatomist), Charles Darwin (a divinity student and naturalist), Robert Chambers (a journalist), William Buckland (a theologian), Roderick Murchison (a soldier and man of leisure), Adam Sedgwick (who, when seeking election to the chair of geology at Cambridge, boasted that he knew nothing of geology), Hugh Miller (a stonemason), John Fleming (a zoologist), and others of like sort. (Morris, 1984, pp. 301-302; emphasis added)

Malcolm Bowden, a British engineer and a long-time student of Darwinian history, reveals the glaring anti-Biblical bias that Charles Lyell kept close to his vest for a long, long time. Remember, it was Lyell who is credited with completing and popularizing the "new" theory. Bowden wrote:

Malcolm Bowden

> That Lyell had produced three large volumes supporting the [uniformitarian] theory certainly suggested that the case was based upon thoroughly researched evidence. However, as will become clear in due course, the case actually rested upon

> the imposition on the fossil record of a preconceived idea, any contrary evidence being deliberately ignored or made light of. (Bowden, 1982, p. 23)

Within forty years of the publication of Hutton's two-volume work, Lyell had managed to create a story about Earth that congealed Hutton's ideas into a believable, understandable theory. Lyell was a trained British attorney, but he soon left his law profession to pursue his ideas about the earth.

By 1830, Lyell had written and published the first volume of his three-volume work entitled *The Principles of Geology* and, uniquely, a copy of this first volume was given to young Charles Darwin by John Henslow (one of Darwin's professors at Cambridge) before he left England on his historical round-the-world-trip aboard the HMS Beagle in 1831. The ideas presented in this book so influenced Darwin that he arranged to have the other two copies forwarded to him upon their completion, which he received during this trip—a trip that lasted nearly five years, from 1831 to 1836.

Lyell's ultimate influence of Darwin's thoughts is borne out in his own writing. Darwin confessed that his thoughts and ideas **"came half out of Sir Charles Lyell's brain"** (Hitching, 1982, p. 227; emphasis added). There is no question that Lyell greatly influenced Darwin. This is a significant point and one that we must not forget! But more important than this was the quietly held motive that provided Lyell with the impetus for his writing and speaking. Historian Dr. Gertrude Himmelfarb, in her masterful biography of Darwin's life and legacy, referred to a piece of correspondence from Darwin to his son, George, that aids our understanding of Lyell's true interest and motive in uniformitarianism:

> **Lyell is most firmly convinced that he has shaken the faith in the Deluge** (the Genesis Flood) far more efficiently by never having

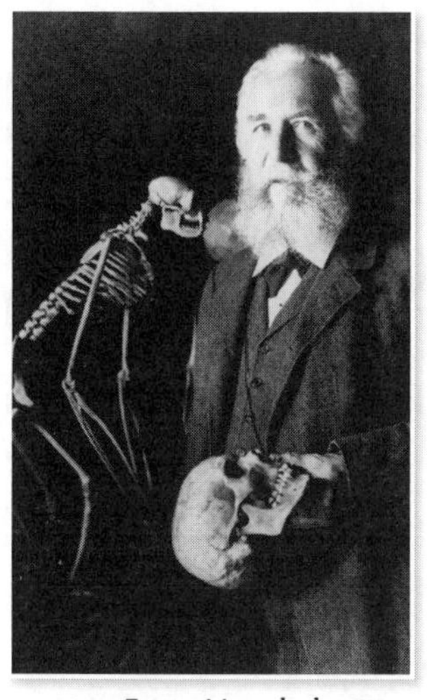

Ernst Haeckel

said a word against the Bible, than if he had acted otherwise. (Himmelfarb, 1996, p. 387; emphasis added)

Lyell did not like Genesis, or its conclusions. It is difficult to believe that Lyell had this agenda so carefully concealed, but he did! Further, it is virtually certain that he encouraged Darwin to continue his work even though he publicly rejected evolutionism early on in order to provide legitimacy for his own theory of the earth. This is made very clear in a letter Lyell wrote to Ernst Haeckel in November of 1868 in which he told Haeckel that six editions of his *Principles* had prepared the way for Darwin (Lyell, 1881, p. 436). However, in keeping with the true nature of the War between good and evil, so that we do not lose sight of how this aspect of nineteenth century history played into this particular battle of that War, Dr. Henry Morris writes:

> ...despite his profound influence on Darwin, Lyell vacillated a long time before becoming an open disciple of Darwinism. Himmelfarb argues persuasively that this was for political reasons, Lyell wanting to retain favor with the theological, political and scientific establishment, which were still committed at least nominally to creationism. It was far safer, and in the long run more effective, to get uniformitarianism and the great age of the earth firmly established before openly endorsing evolutionism. (Morris, 1989, p. 164)

In an 1830 letter that Lyell wrote to George P. Scrope, one of his personal disciples, he openly reveals this wicked nature of his scheme:

> If we don't irritate, which I fear that we may...we shall carry all with us. If you don't triumph over them, but compliment the liberality and candor of the present age, **the bishops and the enlightened saints will join us in despising both the ancient and the modern physico-theologians** ("physico" is a coined expression referring to the Flood)...I conceived the idea five or six years ago, **that if ever Mosaic geology could be set down without giving offence, it would be in an historical sketch.** (Lyell, 1881, pp. 270-271; emphasis added)

It is this kind of subterfuge which is the hallmark of Lyell's life and of modern evolutionists. He definitely wanted to destroy the timeline presented in Genesis. He despised Bishop James Ussher's chronology, which was based on a literal accounting of Genesis 1-12, because he discounted Biblical authority in general. I again appeal to Dr. Henry Morris, whose assessment of the above letter truly explains Lyell's ultimate agenda:

> Lyell did indeed devote a great many pages in the opening section of his *Principles of Geology* to his "historical sketch" of the development of geology, **thus giving him an opportunity to ridicule over and over again what he called the "Mosaic geology" of his predecessors—by which he meant the Biblical chronology and especially the worldwide flood of the Bible and its geological significance.** He did this subtly, however, never referring to the Bible directly and never advocating evolution...**Lyell's dominating motivation, though he was always careful not to express it publicly, was his desire to undermine the authority of the Bible.** He could do this most effectively by first undercutting God's supernatural power and His judgment on sin. (Morris, 1989, p. 164; emphasis added).

Once Lyell completed these history-making volumes (1833), it seemed the Western world was ready for the transition from Biblical creation to evolutionary naturalism. Thus, once again Enlightenment modernism, driven by the idea that man's reason trumps the Bible, won the day and permitted Lyellian uniformitarianism to become the established model of Earth's history for the next 100 or so years (from the mid-nineteenth century to the mid-twentieth century, at least).

Uniformitarian geology is the view of Earth that has been rigorously propagated throughout the Western world since Lyell. Its chief feature—being millions and millions of years of Earth history—has been believed, taught, and maintained to this very day. It is fascinating how this historical scenario is so very reminiscent of the Apostle Paul's warning in Romans 1:25.

I find it incredibly interesting that Alex Marton wrote these salient comments:

> In the 18th century, the winds of democracy from America and the attacks of thinkers like Locke and Rousseau, among others, questioned the Monarchy as a natural form of government. **Liberalism was moving and its method was to go after Biblical Geology (attempting to discredit the global Flood, and thus the Bible),** in order to disarm the Monarchists...What the liberal middle class wanted was reform in Parliament, but traditional theological doctrine stood in the way. Paley's *Natural Theology* claimed that sovereignty descended from God to the King... There was only one way to reform Parliament, and that was to destroy Paley's *Natural Theology*—**and the only way to do that was to discredit the catastrophist notions of its religious defenders who sought to reconcile the geological evidence with the (Flood) story of Genesis**...a young Whig lawyer named Charles Lyell decided to take a novel approach: in his *Principles of Geology*, **he argued against the catastrophists by saying that the diluvial theory (Flood theory) was, in effect, mythology, and that it stood in the way of progress in geology...**After some early skirmishes, Darwin's "theory of evolution" won the day—**a mechanistic theory of evolution subservient to and dependent upon geological uniformitarianism.** (Marton, 1985, pp. 12-13; emphasis added; Marton was a disciple

of Immanuel Velikovsky, who was himself a non-Biblical catastrophist during the middle of the twentieth century—the zenith for the popularity of uniformitarian geology in America.)

NOTE TO THE READER: *A Whig was a member or supporter of a British political group of the 18th and early 19th centuries seeking to limit the authority of the Monarch and increase the strength of Parliament.*

Marton's observations are significant even though he refers to other historical facts that we cannot address in this particular context. He, nevertheless, shows the clandestine nature and inherent evil associated with the overthrow of the Biblical history of Earth in the last 150 years. This should give the reader some comprehensive insight into the insidious and multifaceted extravagance and subtle intrigue which is associated with the War of the Ages.

Even though there are many academic reasons that have led to the professional abandonment by many secular scientists of Lyell's theory today, you can still find geological uniformitarianism being taught in most state-sponsored high schools, colleges, and universities. Sadly, Lyell's theory of Earth has successfully served its intended purpose in the age-long battle between good and evil. It gave credence to Darwinism, and it gave academic sanction to a view of Earth that effectively denied the history and accuracy of Genesis. If the Genesis record of creation and the great Flood is not true as written, then we have no historical basis upon which to contend for the authority of the rest of the Bible, or for the advent and absolute deity of Jesus Christ.

The sad reality in all this was that it motivated a wholesale abandonment of the Biblical record by the conservative theologians in America in the latter part of the nineteenth century. This gave deep inroads into the hearts of many Americans and left them without any effective resistance or argumentation to refute this godless model of origins. This is not to say there were no apologists during this time—there were—but it was too little, too late. By the time Darwin had reached South America

Charles Darwin

on his trip around the world, he had already begun applying Lyell's theory to biology. Thus, Lyell's theory of geology laid the foundation stones for the development of Darwin's model of biology with its millions and millions of years.

The Biblical record was abandoned because "science" had discovered another history of life which "they" contended could be "observed." This, in consort with the other corollaries of Darwinism, has presented Biblical Christianity with its most formidable foe over the last century or so. Dr. Gary North recently wrote in his E-newsletter for the Economic Edge that "Marx is now passe. Freud is suspect...[and] Darwinism reigns almost supreme in every academic discipline..." (Retrieved on August 10, 2009, from http://www.garynorth.com/freebooks/docs/a_pdfs/newslet/position/9405.pdf)

5

The Geological Age System

What about the geological column? It seems to show a historical order of developing life forms, simple to complex, as you proceed from the bottom to the top of the column. But is this an accurate interpretation of the rock record? Good question!

Dr. Gary Parker has provided a great answer to this question. He wrote:

> ...the geological column is an idea, not an actual series of rock layers. Nowhere do we find the complete sequence [especially as they appear in textbooks]. Even the walls of the Grand Canyon include only five [of these] major systems (one, five, six and seven, with small portions here and there of the fourth system, the Devonian)...[However] the geological column does represent a tendency for fossils to be found in groups and for these groups to be found in a certain vertical order. (Morris & Parker, 1987, p. 163)

Dr. Parker continues by saying that Cambrian trilobite fossils and Cretaceous dinosaur fossils are not usually found together. So, why are trilobite fossils and dinosaur fossils not found in the same rock stratum?

For evolutionary theory, the answer seems quite simple. Trilobites became extinct long before the dinosaurs evolved and came on the scene. This conclusion is easy to demonstrate, they believe, by the respective locations in which each of these creatures is found in the rock record. This response is based on the assumption that evolution is true and that evolutionary chronology, as applied to the rock record, is also true. However, we must remember that even if trilobites and dinosaurs were common life forms alive today, they would still not be seen together because their respective habitats were completely different. They lived in different ecological niches. Trilobites lived on the bottom of the ocean, while dinosaurs, for the most part, lived on dry land.

For the creationist, this is a key to understanding the Biblical prediction concerning trilobites and dinosaurs. The sedimentary rock layers and the fossils found in them are not explained by Bible believing creationists to be representative of millions of years of Earth history. For the most part, they are viewed as the products of a worldwide, catastrophic event—the Flood—an event that the Bible teaches took place just 4,300 years ago. And because all animals were created within 48 hours of each other (day five and day six of the creation week to be exact), the Bible seems to teach that their descendants were buried and were fossilized by the Genesis Flood. So, what we observe in the fossil record, for the most part, is nothing more than the geological record of this event.

Accordingly, the Bible predicts the difference in the location of ocean bottom fossils and dry land fossils, such as trilobites and dinosaurs, to be nothing more than habitat zones. Trilobites are found at the bottom of the sedimentary rock layers because that is where they lived. Dinosaurs are found mid-way in the rock record because that is where they lived. Their respective position in the rocks is nothing more than evidence of their own unique habitat or environment in which they lived, were buried, and fossilized. Dr. Parker corroborates this idea:

> After all, even if trilobites and dinosaurs were alive today, they still wouldn't be found together. Why?...Because they lived in different ecological zones. Dinosaurs are land animals, but trilobites are bottom-dwelling sea creatures. According to creationists, the geological systems represent different ecological zones, the buried remains of plants and animals that once lived together in the same environment. A walk through the Grand Canyon, then, is not like a walk through evolutionary time; instead, **it's like a walk from the bottom of the ocean, across the tidal zone, over the shore, across the lowlands, and on into the upland regions.** (Morris & Parker, 1987, p. 165; emphasis added).

Based on the clear predictions of the Genesis Flood model, creationists believe that fossils of particular animals generally appear in the rock record according to their unique habitat, with the exception of those fossils considered disordered or anomalous. Anomalous fossils are a potentially serious problem for evolutionists because they cannot offer an explanation for this disorder. Creationists, on the other hand, can offer a predictable model for the observed irregularities and misplacement of certain fossil specimens. I believe the disorder is directly due to the violent, convulsive nature of the Genesis Flood itself.

It is only reasonable to expect that most creatures whose habitat was the ocean bottom would be fossilized in situ (that is, in their natural location, or habitat). Therefore, we would expect creatures such as trilobites, clams, snails, crinoids, etc. to be buried at or near the bottom of the geological record. Tidal creatures such as fish, squid, octopi, etc. should appear next, and shoreline creatures should appear next (including certain crustaceans, insects, a few amphibians and reptiles, etc.), and so forth. Thus, the supposed geological column found in textbooks is simply a legitimate way in which we can think about the Flood and its influence on rocks and fossils.

To the evolutionist's chagrin, the rock record **does not** exhibit the simple-to-complex evolutionary development of life as they claim. It essentially shows, I think, the results of divine judgment causing the sudden death of the animals and plants that were created in genetically controlled kinds that lived together and were buried together.

So, where did the evolutionary explanation of the rocks and fossils originate? Perhaps the most significant extension, or the greatest effect, of Lyellian uniformitarianism has been the development and propagation of the "geological age system." This naturalistic model for the formation of the rocks and fossils adversely affects one's interpretation of the same rocks and fossils because it gives a forceful, pictorial image that is anti-Biblical in purpose and effect.

What is the true nature of this thing called the rock record? Does it really exist? If it does, what does it tell us? Is it free from human interpretation? And how does it affect my understanding of dinosaurs?

There should be no one who has not seen the standard, textbook portrayal of the evolutionary view of this supposed column. Every textbook of Earth Science from sixth grade through university level geology presents this image of the supposed vertical arrangement of sedimentary layers that is divided into four essential categories called Eras.

What you observe in this naturalistic generalization is a pictorial arrangement of horizontal layers, each with an assigned name and range of age. There is also an accompanying visual representation of many different creatures correlating life at that time and stage in the evolutionary history of Earth. Particular fossils are identified in each geological system called index fossils. This pedagogic image forms the backbone of uniformitarian historical geology—a significant piece of the evolutionary story about Earth. The explanation of this chart goes something like this...

THE UNIFORMITARIAN GEOLOGICAL AGE SYSTEM

Uniformitarian historical geology is best defined by this chart which presents Earth's topographical story from bottom to top. The first Era of time is called the **Precambrian Era** (before the Cambrian), and is generally said to represent a span of time from 3.5 billion to 580 million years ago (you will find that different textbooks present a similar range of these ages). It is in this Era that evolutionists believe the first unicellular life evolved. They call this life proto-life. Thus, these supposed ancient, single-celled creatures are often referred to as "primitive," or "primordial." These are some of the many terms provided by the evolutionary illusion.

NOTE TO THE READER: When discussing geological ages, the ages (or years) are usually considered in order of oldest to youngest.

Next is the **Paleozoic Era**—a derivative of two Greek words, *paleo* meaning ancient and *zoic* meaning life. Generally, this Era is said to represent a span of time from about 580 million years to 245 million years ago and represents the beginning of multicellular life as well. For the first 200 million years or so of this Era, many evolutionists believe that life was confined to the oceans. However, it is noteworthy that evolutionists postulate terrestrial animal life also emerged during this Era sometime during the Devonian Period, and that terrestrial plant life emerged in the earlier Silurian Period.

A big problem for evolutionists is the sudden burst of highly complex, metazoan (multicellular) life-forms that are commonly found in the first Period of the Paleozoic Era called the Cambrian Period. The difficulty is caused by the fact that transitional precursor fossils have not been found explaining the sudden appearance of all the fully developed phyla that are abundant in these layers. This is often referred to by evolutionists as the Cambrian Explosion—more on this later.

In the middle of this evolutionary time chart is the **Mesozoic Era**—derived from the Latin *meso* meaning middle and *zoic* meaning life.

This Era is generally said to represent a span of time from about 245 million years ago to 65 million years ago. It is in this Era that we find dinosaur fossils.

Finally, we come to the **Cenozoic Era**—derived from a Greek word *kainos* meaning new or recent and *zoic* meaning life. This Era is generally believed to represent a range of time from about 65 million years to the present.

Eras are divided into Periods

Each of these Eras is divided into Periods. In the Paleozoic (bottom to top), is the Cambrian, Ordovician, Silurian, Devonian, Carboniferous and the Permian Periods.

The Triassic, Jurassic, and the Cretaceous Periods make up the Mesozoic Era. Who is not familiar with Steven Spielberg's *Jurassic Park?* This Era is the so-called "age of the dinosaurs." At the top of the rock record is the Cenozoic Era consisting of the Tertiary and the Quaternary Periods. This Era is believed by evolutionists to be the age of mammals and man.

This time chart is the Holy Grail for uniformitarian historical geology. Thus, it serves the uniformitarian geologist and paleontologist in their godless endeavor of explaining the rocks and fossils without the Creation or the Flood. This means, for example, any fossil found in rock layers associated with, named, or identified with the "Cambrian" is believed to be in the neighborhood of 500 million years of age, etc.

This system further insinuates that life evolved, red in tooth and claw, from simpler organisms and gradually became more complex from bottom to top in a continuous, interconnected ancestry over millions and millions of years with man being on top—man presently being the highest evolved of all the animals. Therefore, it is assumed that man is nothing more than a sophisticated, big-brained, naked ape and, accordingly, has been assigned to the same Order *(Primate)* with modern apes. From this naturalistic history of Earth, then, all evolutionary, illusionary scientisms are birthed, composed, and publicized.

NOTE TO THE READER: *The term "scientism" refers to a form of semantic illusion deliberately used by pedagogic (educational) materialists to give academic ascendancy or superiority to their evolutionary religion. It involves the arbitrary assignment of scientific status to a philosophical or even a metaphysical idea which is designed to provide it with believability and acceptance over other competing ideas. Scientisms are used by evolutionists in many ways, such as "coined" words and phrases, thought-controlling visual design, or guided correlations. Their intent is to give significance and believability to evolutionary conclusions.*

The geological age system is itself such a scientism, and it is because of this that it is so terribly beguiling. It is just another way Satan questions the reality of the Scripture. In this instance, the question is "Did God really say there was a worldwide Flood?"

Here is how this beguilement is accomplished. It is primarily accomplished because the evolutionary interpretation of rocks and fossils is given unquestioned scientific sanction and authority, and it is successfully used to also prove their "goo to you" evolution of life. This supposed credibility of the geological age system nullifies the Biblical narrative of Creation and the Flood causing these events to appear as mere religious sentiments of uninformed Bible believers. To this reality, however, Professor Gareth Nelson with the American Museum of Natural History honestly admits, as far back as the 1970s, that:

> It is a **mistake** to believe that even one fossil species or fossil "group" **can be demonstrated** to have been ancestral to another. **The ancestor-descendant relationship may only be assumed to have existed in the absence of evidence indicating otherwise**...The history of comparative biology teaches us that the search for ancestors

Evolutionary view of rocks and fossils in the Earth's crust

Era	Period	Events
Cenozoic	Quarternary	Evolution of humans
	Tertiary	Mammals diversify
65 MILLION YEARS — KT BOUNDARY*		
Mesozoic	Cretaceous	Extinction of dinosaurs First primates First flowering plants
	Jurassic	First birds Dinosaurs diversify
	Triassic	First mammals First dinosaurs
Paleozoic	Permian	Major extinctions Reptiles diversify

0 1.8 50 100 150 200 250 300

MILLIONS OF YEARS AGO

*KT is the abbreviation of the German word for Createaceous. The KT Boundary represents the supposed evolutionary point at which dinosaurs became extinct.

often referred to as the Geological Column

Era	Period		Events
Paleozoic	Carboniferous	Pennsylvanian Mississippian	First reptiles Scale trees Seed ferns
	Devonian		First amphibians Jawed fishes diversify
	Silurian		First vascular land plants
	Ordovician		Sudden diversification of metazoan families
	Cambrian		First fishes First chordates
Late Proterozoic			First skeletal elements First soft-bodied metazoans First animal traces

350 400 450 500 550 600

MILLIONS OF YEARS AGO

> is doomed to ultimate failure, thus, with respect to its principal objective; **this search is an exercise in futility.** Increased knowledge of suggested "ancestors" usually shows them to be too specialized to have been ancestors of anything else...In contrast to what is usually stated, therefore, a more complex sample of the fossil record in itself would only complicate the problem of assessing the interrelationships of the fossil species...That a known fossil or recent species, or higher taxonomic group, however primitive it might appear, is an actual ancestor of some other species or group, **is an assumption scientifically unjustifiable,** for science never can simply assume that which it has the responsibility to demonstrate... It is the burden of each of us to demonstrate the reasonableness of any hypothesis we might care to erect about ancestral conditions, keeping in mind that we have no ancestors alive today, that in all probability such ancestors have been dead for many tens or hundreds of millions of years, **and even in the fossil record they are not accessible to us.** (Nelson, 1971, pp. 18-30; emphasis added)

Nevertheless, secular scientists and state schools continue to blatantly present evolution as an irrefutable, demonstrable fact while, at the same time, presenting the Biblical record as being merely religious and having nothing intelligent to add to the discussion. Thus, the sacred reality of the "divine" in man's original creation is continually obfuscated by relegating man to being descended from animal ancestry, which makes him a mere relative to the rest of the animals and nothing more! It is revisionism with a vengeance! It removes God from history. As a result, He can no longer be the Alpha and the Omega; thus, His first advent and His promised return are both thoroughly negated. Furthermore, it alleges that evolution is continuing, and will continue, well into the future even though they admit they do not have the foggiest notion where it's going, or what it will produce. It is the most formidable enemy of Biblical truth in the last 150 years.

The sad fact in all this is that very few see the deliberate semantic illusion in the evolutionary presentation. Dr. Harold R. Booher has thoroughly assessed the subtly of this academic strategy as it relates to the geological age system:

> The method used by scientists to determine how fossils from all over the world can be assigned to successive strata, is to visualize a continuous geological column and assign fossils to various positions on the column by correlational techniques. By drawing horizontal lines on a page, separating them by various distances and giving the spaces between lines names representing different time periods, the image of a geological column is created. For any piece of the column, say the Devonian or the Cretaceous Period, the correlational techniques attempt to consider geological characteristics of actual strata which might be indicative of the period, but most importantly are the fossils themselves. For example, the Devonian...sometimes called the age of fishes, is assigned a time (span) of about 60 million years. Characteristic fossils assigned (often referred to as "index fossils") are certain jawless fishes, early jawed fishes and by the end of the Period, lungfishes. Anywhere in the world where similar fish and other associated fossils are found, their encasing geological strata could be (and most often is) labeled Devonian. Although correlational techniques can be quite sophisticated and often show great consistencies in geological formations housing fossils, any actual column section is generally small. In fact, there is no such thing as a complete geological column anywhere on earth. (Booher, 1998, p. 65)

In fact, geology Professor E.M. Spieker of Ohio State University, said:

> ...how many geologists have pondered the fact that lying on the crystalline basement are found from place to place not merely Cambrian, **but rocks of all ages** [Note: When using the term "basement," Spieker is referring to the underlying igneous and metamorphic rocks—most often the granites deep in the earth

> which form the foundations of the Earth.] And what, essentially, is this time scale?—On what criteria does it rest? When all is winnowed out and the grain reclaimed from the chaff, it is certain that the grain in the product is mainly the paleontologic record and highly likely that the physical evidence is the chaff. (Spieker, 1956, pp. 1805-1806; emphasis added)

What Dr. Spieker has disclosed has been common knowledge for years, a fact that is in total agreement with Professor John Woodmorappe (see quote on p. 22). There simply isn't a continuous, systematic, global sequencing of rock layers as has been suggested by the evolutionists anywhere in Earth.

For this study, it is important to understand what evolutionists think caused dinosaurs to become extinct that marked the beginning of the Cenozoic Era about 60 to 64 million years ago. They are convinced they can prove this by evidence in the rock record. Moreover, they are just as convinced that humans first appeared approximately one to four million years ago—about 60 or so millions years after dinosaurs ceased to exist. But can these assumptions really be proven?

Other significant scientisms are words and phrases such as "molecules to man," "particles to people," "common ancestor," "family tree," "primitive," "advanced," "ontogeny recapitulates phylogeny," and so forth. There are literally scores (maybe hundreds) of these naturalistic clichés and iconic terms that show up in scientific articles and textbooks that are hardwired into the literature as support for the evolutionary story. All scientisms find their origin and support in the belief that evolution actually happened and that, for some, it is still happening.

Dr. Booher accurately assesses this situation for what it really is:

> Whether they realized it or not, teachers of evolution rely heavily on visual aids designed to instill "good Gestalt" (perception forming content) in an **illusionary fashion.** They depend on imaginary

> columns connecting geological formations, fruit laden phylogenetic trees, embryological shadows, and flesh covered bone scraps to make Hyde-Jekyll changes from *Ramapithecus* to *Homo sapiens* a reality...The rocks show catastrophe and discontinuity throughout ages past. This perception held until invisible underlying factors integrated them into a larger smoother picture...Science should help us to sort out what is true and what merely gives a pleasant but flawed feeling of knowing. **But it is extremely difficult to unseat our most prized icons at the highest levels of explanation if they are continually reinforced at the perceptual level with satisfying illusions.** (Booher, 1998, pp. 52-53; emphasis added)

The use of "scientisms" has become so widespread and popular that they have become the "icons" of evolution. And for most laymen and scientists, these "icons" are themselves the unquestionable evidence of evolution. Ten of these "icons" have been elevated to the status without which evolution cannot exist or be understood. (Wells, 2002, pp. 1-8) However, what I want the reader to understand is that much of the advertised support for these "coined" and "contrived" expressions is transferred from the evolutionary interpretation of rocks and fossils. Therefore, the evolutionary view of dinosaurs is simply the naturalistic presupposition that has been concluded from an evolutionary interpretation of the rocks and fossils—all based on uniformitarian assumptions. Dare I repeat this for emphasis! I am telling you without a doubt, that evolutionists offer evidence for their view of dinosaurs (and for all other of their origins conclusions) from their evolutionary interpretation of the evidence which is then presented as "scientific" evidence. It is interpretation presented as evidence!

If dinosaur fossils begin appearing in rock layers labeled Triassic, which is the assigned name of the first rocks containing dinosaurs (itself a tautology), evolutionists believe this is the beginning of the "age of dinosaurs." As we have learned, the Triassic Period is part of the Mesozoic system, and because these layers of rock have been arbitrarily assigned ages between 245 million to 65 million years, evolutionists conclude dinosaurs lived and died millions of years before man. Moreover, based on this model of Earth, they also conclude that dinosaurs are the result of thousands to millions of biological transitions that began with some single-celled organism hundreds to thousands of millions of years ago—the common ancestor believed to be the prototype to all biological life.

But the evolutionary process and God-defying story does not stop with dinosaurs. For many evolutionists, these unique reptiles eventually became birds. From there, they allege dinosaurs became extinct due to some "catastrophe" at the close of the Cretaceous Period because they indicate dinosaurs don't show up in the rock layers considered younger than Cretaceous. Hence, dinosaurs are believed to have existed 240 million years ago and became extinct 60 to 65 million ago. So, this puts dinosaur extinction, in the evolutionary scheme of things, as I have said, 60 to 65 million years ahead of the appearance of man.

But how do they know this? The truth of the matter is, they don't know this, not for sure; but they definitely believe this. That is the point: Evolution is a belief system about origins that includes definitive ideas about dinosaurs. It is not scientific reality. It is their belief system! If they want to believe these God-rejecting ideas, it is okay with me. But what is NOT okay with me is their academic presumption that their view is scientific and my view is only religious! As such, they contend their view is the only accredited view which should be taught to our sons and daughters. The truth is their view is just as religious as mine.

The problem with this view is twofold. It negates Biblical creation as a possible explanation for origins, and it "willfully ignores" the Genesis Flood as a historical possibility. Because it does, it obviously skews their interpretation of how the dinosaurs got here in the first place, and it causes problems understanding the rocks and fossils. Worse yet, it denies the sovereignty of God as the supreme Ruler and Judge of the universe.

Thus, they interpret the sedimentary layers of the earth to have been deposited gradually over millions of years and the dinosaurs to have been produced by gradual evolutionary processes. Regarding their notion of the global extinction of dinosaurs, as we will see, they are not sure about that either.

6

What if there was a Global Flood?

Dr. John Morris

Some years ago I heard Dr. John Morris, Ph.D. in Geological Engineering from the University of Oklahoma and President of the Institute for Creation Research, make an exceedingly important statement. He said that the Genesis Flood was the bottom line in the creation/evolution controversy because if there had been a worldwide, mountain-covering Flood as the Bible says there was, it would have laid down most of Earth's rocks and fossils. Moreover, he stated that the majority of this geological work, if not all of it, would have happened in just a little more than one year (about 370 days) and not millions of years as evolutionists assert.

In other words, the Genesis Flood is the mechanism that would have deposited most of the present sedimentary layers that comprise the geological record. If this is true, then all the fossilized flora and fauna found in sedimentary rocks must have lived contemporaneously (at the same time) and were killed and buried by the same event, separated only by their habitat. The evidence for this is overwhelming.

I think this view of Earth is plausible for a number of reasons. Genesis 7:11 tells us that the initial force of the Flood caused "all of the fountains of the great deep to be broken up." These powerful forces, near nuclear in power or maybe more than nuclear in power, must have been mind-boggling. The Genesis Flood was a terrible judgment. Nothing before or since even comes close to this event in power and intensity. There is no way we can fully understand the massive intensities and process rates that were associated with this one hydrological event. In the first place, the Genesis text tells us that the Flood was global in extent; and, secondly, it initially included the unleashing of "all" the fountains of the great deep.

The Global Nature of the Flood

I know there are those who quibble about the global nature of the Genesis Flood. Nevertheless, when it is all said and done, it is just that, a mere quibble. It is nothing more than a human argument that has a God-defying motive. Not even a blind man on a galloping horse can miss the clear global intent written into the content of Genesis 6, 7, and 8. To argue for a local or regional application of these verses is nonsense!

Consider why God commanded Noah to prepare the Ark to receive two of every kind of land animal created on day six, plus seven of the winged animals and clean animals. If the Flood was not global, what purpose was there for God demanding Noah and his family, with all these animals, to ride through a "local" or "regional" flood when it would have been much smarter and easier for them to move? Just across the mountain would have been more than sufficient for them to be safe.

The same is true for the "rainbow covenant." Carefully read Genesis 9: 8-17, answer the following questions, and give particular consideration to my conclusions for the fourth and fifth questions:

1. What was the reason God gave for His use of the "rainbow?" *(Answer: Genesis 9:13-15)*

2. Have there been any local or regional floods on the Earth since the Genesis Flood? *(Answer: Yes! There have been multitudes. The promise did not say there would never again be flooding waters on the earth. It said God would never again destroy the earth in this manner. Genesis 9:11, 15)*

3. Does God lie? *(Answer: No! See Numbers 23:19; I Samuel 15:29; Titus 1:2; Hebrews 6:18).* The Hebrew and Greek words used in these verses translated "lie" all connote, in the order presented above:
   - that God cannot lie or deceive,
   - that God cannot cheat or deal falsely,
   - that God is free of falsehood, and
   - that God cannot deceive by lying.

4. Since the "rainbow covenant" was a sign to man that God would never destroy the earth in this same manner, and since there have been many local or regional floods since the days of Noah, and since it is impossible for God to lie, it stands that the Genesis Flood was more than a local or regional Flood. It must have been worldwide, which is the clear intent of the context (compare the context and meaning of the wording in Genesis chapters 6:6, 11-13, 17; 7:3-4, 6, 10-11, 15, 17-20, 22-24; 8:3, 9, 13, 17, 21).

5. The reader must not forget that Jesus used these conditions surrounding the Flood as a warning concerning His second coming. He said in Matthew 24:36-37:

> But of that day and hour no one knows, no, not even the angels of heaven, but the Father only. **But as the days of Noah were, so also will the coming of the Son of Man be.** (emphasis added) (NKJV)

Will people be saved and go to heaven from all over the world? Of course they will. Is the rapture of the Church a global event? Of course it is. Now those people who contend that the Genesis Flood is local or regional must, it seems to me, also be prepared to limit the circumstances or the "signs" preceding the Second Coming of Jesus Christ to be limited to a local or regional area of the world. Since they place limits on the Genesis Flood, it follows that their thinking also places limits on the signs of the Second Coming of Jesus, as well. Of course the Flood of Noah must have been global because the degree and extent of the conditions associated with Noah's Flood will be the same degree and extent of the conditions associated with Christ's Second Coming. Let's continue.

"All" the Fountains of the Great Deep

The energy that drove this year-long hydrological cataclysm seems to have been initiated by global earthquakes and volcanism that caused unbelievable erosion and re-deposition of crustal material (also called diastrophism, meaning processes of deformity in Earth's crust that caused its present continents, ocean basins, mountains, etc.). These global forces exerted in the earth were definitely powerful and inexplicably awesome. It is staggering to consider the potential dynamics contained in the word "all" used in this context (Genesis 7:11). If the Flood was global, and it obviously was, this means the earthquakes and the volcanoes that were associated with the "breaking up of the great deep" were also global. This is mind-boggling!

It is impossible for us to imagine the incredible forces unleashed on Earth when all the volcanoes and earthquakes of the earth were detonated simultaneously. (Note: Dr. John Baumgardner, a leading geophysicist in the United States, has produced a Biblically compatible model called

Catastrophic Plate Tectonics that offers a possible scenario for the forces of this Genesis Flood event. Dr. Walter Brown, former Physics Professor with the Air Force Academy, has developed another salient model for the Genesis Flood called the Hydroplate Theory.)

A brief look at the word translated "fountains" may help us understand, at least a little, the massive extent of this disaster. This word (the Hebrew *ma'yān*), simply means "spring, or water." These "fountains" must have been composed of great volumes of water, and probably other liquefied substances such as magma and brines which were all stored in Earth's crust.

Some massive physical force—maybe an asteroid impact, or it could simply have been the Word of the Lord—caused a general and instantaneous disruption in the integrity of Earth's crust. This disruption gave rise to a total breaching of crustal cover along thousands of miles of fault lines or breaks. Much of this "breaking up" must have been sub-oceanic. The *Theological Workbook of the Old Testament* says of the word "fountains:" "In Genesis 7:11, presumably the reference is to sub-oceanic sources" (Harris, Archer, Waltke, 1980, p. 1614).

This probably released untold amounts of highly pressurized water, together with other liquefied substances, that caused immense volumes of rock debris to be suddenly ripped away and hurled horizontally throughout Earth's biosphere and vertically high in Earth's atmosphere. The abundance and force of the crustal debris, water, lava, and pyroclastic material (hot rock) combined with volumes of volcanic gas that caused unimaginable destruction and rearrangement of Earth's surface.

It is of note, that the phrase "breaking up" (the Hebrew word is *baqa,* pronounced baw-kah, meaning cleave, rend, rip, split or break open or through) is also used in Numbers 16:32. This verse reads, "...and the earth **opened** its mouth and swallowed them up" (emphasis added) (referring to the divine judgment of Korah and his company because they had rebelled against Moses). This helps us understand the devastating nature contained in the in the phrase "breaking up of the great deep" (Genesis 7:11).

You can quickly perceive at this juncture, that the antediluvian people were in desperate trouble and this vast catastrophe was just beginning. Once these initial forces had begun, they were rapidly followed by an overwhelming hydrological event lasting 150 days. Both the immediate and residual forces and processes caused by such an event would destroy all air-breathing life not in the Ark of Noah. It would have rapidly laid down widespread sedimentary layers, thousands of feet thick that quickly formed fossils of plant, insect, and animal life, including dinosaurs. It would have formed coal and petroleum from the deposited plant and animal material, it would have petrified wood, and it would have caused the formation of solution caves (caves formed essentially from sedimentary rock).

The residual effects of the Flood would have caused the formation of mountains and canyons. It would have continued crustal movements which formed Earth's present continents. It would have set in order the physical and atmospheric conditions that ultimately caused an epic ice age following the Flood year. All of these physical changes, and more, in and on the earth would adequately explain Earth's present topographical features and conditions.

All land-dwelling, air-breathing animals (and this would include all reptiles, many of which we now call dinosaurs) that were not on the Ark were destroyed in this great Flood. Some were quickly buried in the sediments of the Flood and were fossilized. Others, I believe, simply decomposed because they were not being buried significantly enough, or not at all, to become fossils. Some of the fossilized bones we presently discover were, it seems, formed in the very spot where they were originally buried. Others, however, seem to have been transported by the waters of the Flood long distances from their original habitat, maybe even deposited, fossilized, unearthed, moved and then re-deposited.

If this Biblical Flood analysis is true, and there is certainly much evidence in support of its veracity, then Dr. John Morris' beginning assessment is right, and the millions of years in the evolutionary story vanish before the authority of the Word of God. This means that the Genesis Flood predicts that the vast majority of this entire rock column (and maybe all of it) was laid down by this global event in about one year—about 4,300 years ago. In other words, the Genesis Flood is, indeed, the bottom line in the origins controversy, not just for our understanding of dinosaur fossils.

Think about this. If most of the disputants in this debate deliberately ignore or even deny the obvious global, catastrophic nature presently seen in the formations of Earth's crustal surface just because they are emotionally committed to millions of years (and they do ignore and deny the evidence), it is impossible for them to ascertain the accurate cause for these features. This self-inflicted ignorance reveals the power of one's worldview; but, tragically, it also causes them to deny the promise of Christ's second coming (II Peter 3:3-4). This is sad, but what I want the reader to see more than this is the true motive behind their denial; a motive that I believe is closely associated with that of Charles Lyell's motive for writing his three volume set, *Principles of Geology.* It is a very wicked motive that is often hid in the "stuff" of scientific jargon.

Most of the ad hoc assumptions promoted by secular geology are the natural corollaries of uniformitarian thinking, Dr. Carl Dunbar notwithstanding. Moreover, because the secular crowd controls the university system in America, both the accrediting process and the teaching standards, it is their interpretation of the rocks and fossils that have negatively affected the worldview of millions of unsuspecting people concerning Biblical reality for the last hundred years or so.

Interestingly enough, Lyellian uniformitarianism is now beginning to lose favor among many secular scientists. They are simply seeing too many evidences of catastrophism in the geological processes to continue to ignore them. This disengagement from the assumptions and predictions of uniformitarianism has been greatly encouraged by the catastrophic nature and effects of the May 18, 1980, eruption of Mount Saint Helens in the state of Washington.

A younger Charles Lyell

What I want to do now, is to discuss a few of the more obvious of these uniformitarian assumptions that remain and then spend a little time exploring the historical nature of the transition from Lyellian uniformitarianism to what is presently called "neo-catastrophism"—a transition that is taking place today.

NOTE TO THE READER: *Neo-catastrophism is the name given to the new geological explanation of the Earth's crustal features that is held by an ever-growing number of secular scientists. This term is only considered new in relationship to the dominance and popularity of uniformitarianism. In fact, the history of science reveals this growing support of neo-catastrophism is more of a revival of a formerly held idea than the discovery of a new model about the Earth. Many secular scientists are now admitting that most features of the Earth's surface were catastrophically formed, but they continue to deny that this catastrophic mechanism was the Biblical Flood. However, they do not deny the obvious catastrophic nature of the rock record which must have caused particular features—whether they were sedimentary formations, fossils, or caves. Many evolutionists freely admit that certain rock formations indicate a global nature to that formation;*

but they believe there were many such catastrophic episodes throughout Earth history. Thus, they maintain the evolutionary time-frame by insisting that large amounts of time passed before and after each catastrophic event or episode.

7

Glaring Assumptions of Uniformitarianism

The first of these assumptions that I want to consider concerns the vast amount of time needed in order for evolutionism to have the slightest possible chance of facilitating the origin of the universe, life, and man. Without a doubt, the most confusing aspect in secular geology is all of the time that is supposedly justified by this discipline. How do geologists decide the age of a rock and its place in the supposed evolutionary sequence? Does radiometric dating provide any accuracy in dating sedimentary layers? Isn't it a fact that most of the rock column, including much of its tremendous age, was relatively identified long before radiometric processes were developed? Yes, it was. But it is also a fact that the evolutionary age of the earth (and rocks) is growing. In this relationship, I think it is incredibly interesting that Fred Whipple with the Center of Astrophysics in Cambridge, MA, said, "The Earth has aged dramatically during the three centuries of the scientific era. **The estimated age doubled on the average every 16 years since the middle of the 17th century...**" (Whipple, 1978, pp. 1-2; emphasis added).

(Please understand that Whipple is not a creationist; nevertheless, he continued in this article to justify the apparent necessity for this doubling of Earth's age. The point is: The age of the earth must still be increasing!)

To illustrate the nature of this problem, I want to relate to you an experience that happened to me in the late 1990s. I had traveled to Glen Rose, Texas, with my wife and another couple. They had taken us there for a weekend outing in their motor coach. Our designated purpose for making this trip was to view the many dinosaur tracks visible in the Paluxy River basin formed by the Glen Rose limestone. Upon arrival, we found a suitable place to park the motor coach in Dinosaur State Park, which overlooks the famous Paluxy River. Once we were suitably settled, I excused myself to go to the river to view and study the tracks.

NOTE TO THE READER: *If you have never been there, you must go soon because the tracks are quickly eroding, and will probably be completely erased by erosion in another fifteen to twenty years.*

While on site, time passed swiftly, and it suddenly dawned on me that I had been in the river three or four hours. Becoming aware of the time, I felt it necessary to return to the motor home as it was about lunchtime. I climbed out of the river and began my walk back across the park when I noticed there were several 15-passenger vans lined up in the parking lot that serves an area where the park authorities had placed two large, somewhat exaggerated, models that represent the dinosaurs whose tracks are found in the riverbed. As I came closer to this area, I saw that the vans were from the University of Texas, Austin. I walked into the parking area and realized there were forty or fifty undergraduate students there on a geological field trip. As I drew closer to the group, it became obvious that they were listening to their professor, so I stopped for a minute so I could listen. I became so intrigued with what she was saying that I eased into the group. I had totally forgotten lunch! About ten minutes after I arrived, the professor asked the students if they had any questions. There was silence, so I raised my hand and asked if I could ask a question. The professor said that would be just fine (I suppose she wanted someone to ask a question). I asked the age of the Glen Rose limestone in which all of the dinosaur tracks are found. The professor said, "That's easy enough. The Glen Rose limestone in which the tracks are found is between 90 and 110 million years old." I responded, "Wow, that's old!", and the professor continued teaching. After another ten minutes or so, the professor again asked if anyone had a question. And again, the students were deathly silent; so, once again, I raised my hand and was recognized. I asked the professor if I was interrupting the class and she said, "No, not at all." This time I asked how she knew the Glen Rose limestone was so old. She said they knew this because the rocks were dated by the fossils that were contained in them. I was a little shocked at this, however, I said, "Oh, that's interesting!"

The professor taught for a while longer, and again asked for questions. Once again, there was silence (I think the students wanted to get back to Austin because it was Saturday). So I again asked another question. "Since you date the rocks with the fossils they contain, how do you know the age of the fossils?" What the professor said next astounded me. I had heard other creationists mention this in discussion, but I had never heard it for myself. She said, "We date the fossils with the rocks in which they exist." This is classic uniformitarianism! They date the rocks with the fossils, and the fossils with the rocks! The answer is completely tautologous (circular), which, I think, simply tells us they really don't know!

The truth of the matter is, the long ages supposedly vindicated by the rocks and fossils were put in place by the early uniformitarians—circa 1830 to 1860. (Perloff, 1999, p. 153) In order for proponents of this model to maintain any semblance of consistency correlating rock ages with particular fossils, they needed to identify particular fossils to be what they called "index fossils." By definition an "index fossil" is:

> ...the fossil remains of an organism that lived in a particular geologic age, used to identify or date the rock or strata in which it is found: Good index fossils must be abundant and widely distributed over the earth, and large enough not to be overlooked. (Barnhart, 1986, p. 314).

I have often wondered how, with any certainty, the geologist was able to identify a fossil, its age, and geological origin. They are able to do this because they have developed a correlated series of so-called "index fossils." Of course, the arbitrarily assigned ages of the rocks based on this series of fossils must presume Darwinian evolution to be true.

All "index fossils" are believed to be extinct and are only known from the rock layers in which they are found, or with which they are uniquely associated. Drs. Henry and John Morris corroborate this fact by telling us:

> ...since evolution always proceeds in the same way all over the world at the same time, **index fossils** representing a given stage of evolution are assumed to constitute infallible indicators of the geologic age in which they are found. This makes good sense and would obviously be the best way to determine relative geological age—**if, that is, we knew infallibly that evolution were true.** (Morris, 1996, p. 290; emphasis added)

Therefore, if a fossil is found resting along a road, or upon a hillside, or wherever else it may be found, the evolutionary geologist will quickly tell you that you have found a fossil from the Cambrian, or the Permian, or some other particular geologic age system. They do this because most all fossils found today have already been identified and correlated with one or more of the supposed geological periods. And these geologic periods have been catalogued and advertised from the assumed age of the rock system in which they were identified.

But this interpretation of the rock record is squarely based on Darwinian assumptions. The greatest assumption with the most import is the fact that they believe evolution actually happened in the past and is true, which requires immense amounts of time for the process. However, if Darwin is wrong, the entire historical system is wrong, Lyell and all. The circularity of this position, as mentioned above, is becoming glaringly evident even among many secular scientists. (See O'Rouke's quote below.)

Let me remind the reader of the 1960 comments of evolutionary geologist Carl Dunbar mentioned earlier in which he indicated that the *"doctrine of uniformitarianism demands an immensity of time"* and that *"it has now gained universal acceptance among intelligent and informed people."* This remark, as we shall see, was soon proved to be terribly wrong. Just seventeen years later professional geologist J. E. O'Rourke wrote in the *American Journal of Science:*

> **The intelligent layman has long suspected circular reasoning in the use of rocks to date fossils and fossils to date rocks.** The geologist has never bothered to think of a good reply, feeling that explanations are not worth the trouble as long as the work brings results...These principles have been applied...which starts from a chronology of index fossils, abstracts time units from it, and imposes them on the rocks. Each taxon [a classification name applied to a particular group of plants and animal fossils] represents a definite time unit and so provides an accurate, even infallible date. If you doubt it, bring in a suite of good index fossils, and the specialist, without asking where or in what order they were collected, will lay them out on the table in chronological order. (O'Rourke, 1976, p. 48, 51-52; emphasis added)

Moreover, Assistant Professor of Paleobiology from Kansas State University Ronald R. West, candidly adds:

Contrary to what most scientists write, the fossil record does not support the Darwinian theory of evolution because it is this theory which we use to interpret the fossil record. **By doing so we are guilty of circular reasoning if we then say the fossil record supports the theory.** (West, 1968, p. 216; emphasis added).

A younger Charles Darwin

I again mention for emphasis, that ideas do have consequences; and, in this instance, the consequences are the God-rejecting conclusions of uniformitarian thinking. For example, a considerable amount of the evolutionary time needed to allow the Darwinian strategy to function successfully has been contrived from the Lyellian assumption that "the present is the key to the past," and Dr. A. E. J. Engel, Professor of Geology at Cal Tech, reveals the speculative nature of this contrivance:

> **The fact that the calculated age of the Earth has increased by a factor of roughly 100 between the year 1900 and today**—as the accepted "age" of the Earth has increased from 50 million years in 1900 to at least 4.6 eons [billion years] today—certainly suggests we cloth our current conclusions regarding time and the Earth with humility...We speculate a lot about the first eon or more of earth's history...but in the foreseeable future it will be mostly speculations—essentially geopoetry...The most honest concede freely we know almost nothing of the Earth or its processes. (Engel, 1969, pp. 461, 462, 480)

Isn't it interesting that what we have been told for generations to be scientific fact now turns out to be mere Darwinian speculation; and, to a great extent, this is because they "willfully ignored" the Flood of Genesis. Could it be that all evolutionary thought is just as speculative? I think it is.

Thus, a global catastrophe as described in Genesis 7 and 8 (called the Genesis Flood), along with the many post-flood events (often referred to as residual catastrophism), would adequately explain all of the surface features of Earth we see today. This would include the disappearance of dinosaurs—if, indeed, they are all extinct (more on this later). This means the Lyellian doctrine of uniformitarianism has seriously hampered the proper development of Earth Science. Furthermore, the Biblical Ark/Flood scenario would provide the conditions evolutionists have identified as being associated with the mass extinctions of dinosaurs and other animals while, at the same time, it suggests that varieties of all the original kinds have, since the Flood, been used to repopulate Earth.

Keeping the above information in mind, we ask: Why do most evolutionary scientists ignore the Genesis Flood as a possible explanation of today's Earth? The answer is rather simple. It includes a need for the Creator God of Scripture! Why is this fact such a great problem? To put it briefly, they have been so programmed by the demands of naturalism and materialism to explain their observations that they cannot permit anything Biblical or Supernatural in their thinking. Neither will their humanistic pride permit this kind of holy accountability.

NOTE TO THE READER: *Naturalism is a philosophy that totally rejects any form of supernaturalism to be associated with the physical or the biological world (especially the Creator God of the Bible), primarily because it rejects God's ownership of the universe and the absolute truth of His Word. It became the accepted mechanism for all scientific processes among secular scientists about the time of Darwin. However, the seeds of naturalism can be traced to the Renaissance, but most particularly to the European Enlightenment, and it is the true source for ideas like struggle and "natural selection" in Darwinian thinking.*

Their secular worldview, therefore, simply will not tolerate an explanation of origins to be dependent upon the actions of a world-controlling, sovereign, supernatural intelligence. Thus, for them, any idea that reflects Biblical reality cannot possibly be scientific. Because of this, they abhor faith-based conclusions, notwithstanding their own glaring need for faith in support of their own model. The truth of the matter is, if they ever allow the possibility of a supernatural being in their thinking—specifically the Creator God of Holy Scripture—then they are immediately accountable to Him, and that's an anathema to a secularist. The idea of an Almighty God whom they must honor and obey is far more than their pagan mindset can tolerate.

I think this idea is best captured in a statement by Dr. Richard Lewontin, an evolutionary Marxist and Biologist from Harvard:

> We take the side of science in spite of the patent absurdity of some of its constructs, in spite of its failure to fulfill many of its extravagant promises of health and life, in spite of the tolerance of the scientific community for unsubstantiated just-so stories, because we have a prior commitment, a commitment to materialism. It is not that the methods and institutions of science somehow compel us to accept a material explanation of the phenomenal world, but on the contrary, that we are forced by our...adherence to material causes to create an apparatus of investigation and a set of concepts that produce material explanations, no matter how counter-intuitive, no matter how mystifying to the uninitiated. Moreover, that materialism is an absolute, **for we cannot allow a Divine Foot in the door.** (Lewontin, 1997, p. 31; emphasis added)

Again, we see the use of "science" as a naturalistic replacement for evolution. But more important is the fact that Dr. Lewontin is brutally honest with the evolutionary position; that evolutionists have a deeply instilled commitment to materialism and cannot permit a "divine foot in the door."

8

Cambrian Explosion and Index Fossils

There are two additional facts that must be included in this area of discussion so that the reader will realize the depth of the problem facing present-day uniformitarians. First, I want to touch succinctly upon the implications of the Cambrian Explosion, along with the fact that many forms of life formerly thought to be extinct within the evolutionary scenario have been found alive (including many creatures formerly identified as "index fossils").

The Cambrian Period, as you recall, is the bottom of the geologically identified system called the Paleozoic Era and is assigned an evolutionary age beginning about 500 million years ago. The rock system located just beneath the Cambrian is called the Precambrian, and it is believed to have begun somewhere about 4,500 million years ago. Yes, you read the age accurately. This highly exaggerated number is read 4,500,000,000 and some evolutionists even assign the Precambrian an older beginning.

The natural prediction of Darwinism contends there should be untold numbers of intermediate forms leading up to every new species that appears in the fossil record. Therefore, the fossil record should contain millions times millions of transitional forms (otherwise known as "missing links"). Their absence caused Darwin considerable problem to the point that he wrote a chapter in his *The Origin of Species* under the heading "Difficulties of the Theory." I think it is interesting that he began this chapter (Chapter 6) by saying:

> Long before the reader has arrived at this part of my work, a **crowd of difficulties** will have occurred to him. **Some of them are so serious that to this day I can hardly reflect on them without being in some degree staggered;** but, to the best of my judgment, the greater number are only apparent, and those that are real are not, I think, fatal to the theory.
>
> These difficulties and objections may be classed under the following heads: First, why if species have descended from other species by fine gradations, **do we not everywhere see innumerable transitional forms?** Why is not all nature in confusion, instead of species being, as we see them, well defined? (Darwin, 1998, p. 212; emphasis added)

Darwin recognized the fact that these much needed transitions were missing, and he was keenly aware of the danger their absence caused his theory. However, to be fair to Darwin, he later explained in this same chapter that their absence was due to the youth of geology because only a modicum of fossils had been collected at the time. He indicated that once the fossil collectors had scoured the globe and gathered fossils, the problem of "missing links" would be forever solved. However, to the chagrin of evolutionists, the illusive "missing link" is still missing even though millions of fossils have been found and classified, including at least 250,000 fossils species (Prothero, 2004, p. 18). Yet, among them, there has not been one alleged transitional form that has survived the critical eye of peer review. This is not to say there have not been scores of fossils presented as if they were the long sought for "missing link;" but when all the dust and controversy has settled, this highly sought for status had to be abandoned.

Probably one of the most significant admissions ever made by a professional evolutionist was made in 1979 by Dr. David M. Raup, a geologist at the University of Chicago. He plainly admitted that the needed fossil transitions to prove Darwin's view had not been found as Darwin thought they would: "Well, **we are now about 120 years after Darwin and**...ironically, we have even fewer examples of evolutionary transition than we had in Darwin's time" (Raup, 1979, p. 25; emphasis added).

Of course, Dr. Raup embraces the newly developed evolutionary view of "punctuationism," which is an alternative view developed by Niles Eldridge and Stephen Gould near the mid 1970s. This view developed essentially because of the obvious absence of transitional forms, or any observed progressive order of fossils in the rock record.

NOTE TO THE READER: Punctuated Equilibrium, also known as "punctuationism" is a modification of the Darwinian cause for speciation which argues that "natural selection" acts on a species to keep it stable rather than to alter it. And that evolutionary change happens suddenly, leaving no possible fossil record of the stages through which the organism passed on its way in becoming a new species. Thus, they contend evolution is still true, but the fossil record doesn't show Darwin's bottom line assumption—"missing links."

Dr. Raup wrote, in the same article (above) that:

> Instead of finding the gradual unfolding of life, what geologists of Darwin's time, and geologists of the present day actually find is a highly uneven or jerky record; that is, species appear in the sequence very **suddenly,** show little or no change during their

> existence in the record, then **abruptly** go out of the record. And it is not always clear, in fact it's rarely clear, that the descendants were actually better adapted than their predecessors. In other words, biological improvement is hard to find. (1979, p. 23; emphasis added)

This is a very revealing quote. It clearly demonstrates that evolutionary interpretation is totally colored by their preconceived faith in some kind of naturalistic process. The fact that they can only observe the sudden or abrupt appearance of species can only be explained by some kind of evolutionary mechanism that leaves no evidence of transitional forms. And the fact that they observe the sudden disappearance of this same species requires the extinction of this species without an apparent cause. In both instances, they claim there is no need for transitional forms. This provides them a feasible explanation for the absence of "missing links."

The "punctuationist" idea, crudely stated, is that a mother reptile laid her eggs as many others before her had been doing for millions of years. Only this time, when the eggs hatched, some portion of the hatchlings were distinctly birds, or bird-like with feathers and all. Evolutionist's claim that this happened because hundreds to thousands of undetectable genetic mistakes (genetic copying errors called mutations) have built up over an immense amount of time and suddenly express themselves. This sudden jump from reptile to bird would leave no transitional fossils and is the only explanation evolutionists can give for the obvious absence of transitional forms without violating the "fact" of evolution. The apparent "gap" between the reptile and the bird commonly seen in the fossil record is then explained from this point of view. I know this explanation is a little simplistic and overworked, but it is the essence of the story.

Many evolutionists have embraced this view in the last thirty or so years because the needed missing transitions that Darwin thought would surely be found in the fossil account have not been found, leaving them with no defense against the Biblical creationists. It is interesting that Dr. Raup also said:

> In the years after Darwin, his advocates hoped to find predictable progressions (in the fossils record). In general, these have not been found—yet the optimism has died hard, **and some pure fantasy has crept into textbooks.** (1981, p. 289; emphasis added)

Creationist debaters such as Dr. Duane Gish, Dr. Charles Jackson and others (Dr. Gish participated in over 300 collegiate debates in the latter half of the twentieth century), have seriously impacted the credibility of this evolutionary fallacy to the degree that Dr. Mark Ridley, Zoologist, Oxford University, England, wrote: "In any case, no real evolutionist, whether gradualist or punctuationist, uses the fossil record as evidence in favor of the theory of evolution as opposed to special creation" (Ridley, 1981, p. 831).

Duane Gish

How ironic is this quote! Because for years (nearly 100 years), the evolutionary community, almost to the person, believed the fossil record was the best evidence of evolution. The above position espoused by Dr. Ridley reflects a change in evolutionary thinking. Just twenty years earlier, Dr. Carl Dunbar, geologist from Yale University, said: "Although the comparative study of living animals and plants may give very convincing circumstantial evidence, **fossils provide the only historical, documentary evidence that life has evolved from simpler to more and more complex forms**" (Dunbar, 1960, p. 47; emphasis added).

Dr. Dunbar clearly says that the best evidence for evolution is the fossil record. This was the common belief for most evolutionists during the first sixty to seventy years of the twentieth century. For example, famous scientists like **Pierre Grassé**, the chairman of Evolution at the University of Paris (also called the Sorbonne), and a leading zoologist in Europe, **George Gaylord Simpson,** who received his doctorate from Yale who became a staff member with the American Museum of Natural History in New York City, are two of the well-known scientists to vigorously support this view.

Why do you suppose Dr. Ridley would advocate that the evolutionary crowd should cease using the fossil record as proof of evolution? Because it became continuously clear that what Darwin believed about the fossil record, and what the evolutionary community endorsed, could not be found in that record.

The evolutionary position and explanation about the fossils had to change. In short, this change was facilitated due to two glaring realities: (1) Evolutionists did not find the so-called transitional forms, and (2) There has been a continual, progressive opposition from Biblical creationists over the last fifty to sixty years.

Thus, they saw that the proverbial "walls of Jericho" were about to fall, and as early as 1960, Professor T. Neville George, a paleontologist at Glasgow University, said:

> **There is no need to apologize any longer for the poverty of the fossil record.** In some ways it has become unmanageably rich and discovery is outpacing integration...The fossil record nevertheless continues to be composed mainly of **gaps**...Granted an evolutionary origin of the main groups of animals (as fish, amphibians, reptiles, birds, etc,), and not an act of special creation, the absence of any record whatsoever of a single member of any phyla in the Pre-Cambrian rocks remains as inexplicable on orthodox grounds as it was to Darwin. (George, 1960, pp. 1-5; emphasis added)

Even George Gaylord Simpson admitted:

> In spite of these examples, it remains true, as every paleontologist knows, that most new species, genera, and families, **and that nearly all categories above the level of families, appear in the record suddenly and are not led up to by known, gradual, completely continuous transitional sequences.** (Simpson, 1953, p. 360; emphasis added)

It must be noted by the reader that these so-called "gaps" are nothing more than assumptions of evolutionary preconceptions; they are exactly what a Bible believer would expect to find in the fossil record. Distinct separation between major groups (fish, amphibians, reptiles, birds, etc.) is a predictable fact of the sacred Scriptures.

Drs. Henry and John Morris, tell us:

> ...evolutionists may believe that **gaps** still result from the rarity of fossil deposition and discovery. Or perhaps, as most think, the **gaps** result from spurts of explosive evolution caused by periods of intensified cosmic radiation or something else. The fact is, the **gaps** are still there, and this is a primary prediction of the creation model. (Morris & Morris, 1996, p. 53; emphasis added)

As I see it, when you abandon the Bible, you are headed for dangerous waters and this is exactly the path Darwin took. He did not hesitate to speculate in the last chapter of his famous book:

> Thus from the war of nature, from famine and **death,** the most exalted object which we are capable of conceiving, namely, **the production of higher animals, directly follows.** There is grandeur in this view of life, with its several powers, having been originally breathed by the Creator into a few forms or into one; and that, whilst this planet has gone cycling on according to fixed laws of gravity, **from so simple a beginning endless forms most beautiful and most wonderful have been, and are being evolved.** (Darwin, 1998, pp. 648-649; emphasis added)

In other words, Darwin believed, as do all who embrace his views, that from the very first organism, environmental pressure would stimulate "natural selection" to refine a given organism's ability to compete and survive. From generation to generation, there would be a continuous, incremental progression of more advanced species guided only by natural selection which would ultimately produce survivable organisms. Only these selected organisms would survive and reproduce. This is often referred to as the "survival factor" in evolutionary nomenclature. This "survival factor" would secure the changes necessary for each successive generation to effectively compete for food and habitat. So that over time (evolutionary time), this continuous chain of existence is repeated and species evolve into other species; and each is better adapted for survival than the preceding generation. Ultimately, they are all related to each other and to the original prototype, and we arrive at the level of sophisticated speciation of our present world.

This process is called "descent with modification" and it is the epitome of Charles Darwin's faith. However, not only have **no** transitions been found, but one former evolutionist, the Senior Paleontologist at the British Museum of Natural History, Dr. Colin Patterson, said in 1982: "**No one has ever produced a species by mechanisms of natural selection.** No one has ever gotten near it and most of the current argument in neo-Darwinism is about this question" (retrieved from an interview aired on the British Broadcasting Company, March 4, 1982; emphasis added).

Cracks formed in the evolutionary superstructure because neo-Darwinian scientists like Simpson, Mayr, Dobzhansky, and others simply could not solve the problem of "missing links."

NOTE TO THE READER: *The term neo-Darwinian refers to the group of Darwinians that was forced to combine original Darwinism with the science of genetics soon after the beginning of the twentieth century. This synthesis postulated that evolutionary progression is principally* *due to the operation of natural selection directed by genetic mutations within a population, called micro-mutational transition.*

These men strongly believed that the best evidence for evolution was demonstrated by the fossil record, and they staked their reputation on the idea. However, as World War II came to an end (circa 1945), it soon became recognizable to many secular scientists that their hoped-for transitions in the fossil record would not be found; and, further, that huge gaps existed between major kinds of organisms. They were never able to find fossils expressing incipient structures (i.e., structures coming into being); never were any fossils found that displayed half-feather/half-scale, or half-leg/half-fin, or any other combination of developing characteristics (Morris & Morris, 1996, p.53).

The fossils observed and discussed by Charles Darwin, represented only by animals, plants, and insects that were "well defined," have counterparts still in existence. On the basis of the fossil record, you can only conclude that clams have always been clams, that snails were always snails, and that crinoids were always crinoids, etc. The first time we observed a dog, it was distinctly a dog. Moreover, cats have always been cats. Conspicuously, what we have never observed are "dats" and "cogs"—those in-between creatures necessary to prove evolution.

Did Darwin have a transitional form? No, he did not! There was some quibble in the 1860s about the fossil bird named *Archaeopteryx* being a transition between reptiles and birds, but it turned out to be totally bird—then and now! There is no serious resistance to this fact today. Dr. Alan Feduccia, Professor of Paleobiology at the University of North Carolina, simply said that archaeopteryx could not tell us much about the origin of birds because it was, "**in a modern sense,**" just a bird (Feduccia, 1993, p. 792). Thus, the evolutionary crowd has had to admit:

> The sudden emergence of major adaptive types as seen in the abrupt appearance in the fossil record of families and orders, continue to give trouble. A few paleontologists even today cling to the idea that these **gaps** will be closed by further collecting, but most regard explanation. (Davis, 1949, p. 74; retrieved July 2, 2009, from http://www.pathlights.com/ce_encyclopedia/sci-ev/sci_vs_ev_12b.htm; emphasis added)

So, instead of finding sequential, interconnected transitional forms showing a continuous development from simple to complex in animals and plants, the evolutionists found only "gaps" and "abrupt appearance" in the fossil record.

This sounds very Biblical, doesn't it? If God created all things within their own distinct genetic Kinds, and only permitted them to reproduce within this Kind, you would only expect to find "well defined" creatures in the fossil record. You would expect to see "gaps" and "abrupt appearance." However, the "gaps" are so named by evolutionists because of their assumption of "descent with modification." They are not due to a slow evolutionary progression. The Bible predicts identifiable "gaps" between the "kinds." The idea of "abrupt appearance" is, of course, a prediction of Biblical creation as well. Remember, the fossils can only reflect life as it existed just before the Genesis Flood because it was the Flood that provided the mechanism for most all fossil formation we see today. However, the evolutionist "willfully ignores" the Flood. Therefore, they have no other alternative left than to interpret the fossils from their evolutionary worldview. As a result, they interpret the fossils which exhibit the uniqueness of created kinds with a designed habitat as evidence of "gaps and abrupt appearance. "

That said, consider the following quotes by leading evolutionists:

> Stephen Gould: "All paleontologists know that the fossil record contains precious little in the way of intermediate forms; transitions

Stephen Gould

between major groups are characteristically abrupt." (Gould, 1977, p. 24).

Derek Ager: "It must be significant that nearly all the evolutionary stories I learned as a student…have now been debunked… The point emerges that, If we examine the fossil record in detail, whether at the level of orders or of species, we find—over and over again—not gradual evolution, but sudden explosion of one group at the expense of another." (Ager, 1976, pp. 24, 133).

Lastly, we must briefly return to the testimony of Dr. Colin Patterson who, after a lifetime of evolutionary study, made the following statement in a 1981 interview on the British Broadcasting Co. (BBC):

> All one can learn about the history of life is learned from systematics, from groupings one finds in nature. The rest of it is story-telling of one sort or another. We have access to the **tips of a tree;** the **tree** itself is theory and people who pretend to know about the tree and to describe what went on with it, how the **branches** came off and the **twigs** came off are, I think, **telling stories.** (Patterson, 1981, p. 390; also cited in Gish, 1995, p. 349; emphasis added)

Keep in mind that Patterson served the British Museum of Natural History as a Senior Paleontologist and was a lifelong evolutionist. During his years at the British Museum, he was associated with one of the largest, if not the largest, fossil collections in the world. Moreover, in 1978 he wrote a very thorough book on the subject of evolution in which he invited readers to send him questions. One reader, Luther Sunderland, wrote him and asked why he did not include in his book an example of a transitional form. Patterson wrote back and said that he agreed with the reader regarding the absence of transitions in the fossil record and said that he would have included one in his book, **"if he had known of one, fossil or living."** (Sunderland, 1984, p. 89; emphasis added)

So, what is Dr. Patterson saying? He is affirming exactly what the Bible predicts about the creation of plants, animals, and man. There are no Darwinian-style family trees because no transitional fossils between major groups have ever been collected. All we observe, he explains, are the "**tips** of the branches," which is what the Bible calls "created kinds." Moreover, because the Creator told us in Genesis that the created kinds will only reproduce after their own kind, there can never be transitions between kinds, and this is why the honest evolutionist sees what he refers to as "gaps." There are no real gaps! The so-called "missing link" species never existed. The notion of gaps is just an evolutionary interpretation of the rock record that has been imposed on the evidence. Therefore, it is as Dr. Patterson indicated that anyone who thinks they know how the tree grew (i.e. the tree of life) and how the branches came off (speciation), is simply "story-telling." It is certainly not science!

9

Cambrian Explosion and Living Fossils

There are additional issues of great importance to this subject. The history of science tells us that during the last one hundred years or so, because of the heightened interest in the fossil record, evolutionary scientists discovered an unexpected phenomenon. They found a plethora of fossilized animal phyla all appearing together in the Cambrian rocks. They quickly examined the rocks beneath the Cambrian (called the Precambrian) to see if all of these distinctly modern body forms had any precursor fossils. There were none!

PLEASE NOTE: *The term phyla (singular, phylum) refers to the taxonomic grouping of plants, animals, and insects that are placed just below the kingdom level. Therefore, it is at this level that we find all primary body patterns among animals. Evolutionists include man in the Kingdom Animalia, Phylum Chordata; however, I do not agree with this naturalistic classification for man. Man is not an animal, even though he has a spinal cord! He was created distinctly human in the image and likeness of his Creator God. When believers see this designation, they are often confused regarding man's true identity.*

This sudden burst of life reflected in the fossils of the Cambrian rock layers was soon called the Cambrian explosion and has become another insoluble dilemma for the evolutionist. There is microscopic unicellular life found in the Precambrian (all generally believed to be single-celled prokaryotes, essentially bacteria). But when we progress upwards into the rocks to the Cambrian system, we find all sorts of diversified, multicellular life. It seems from the fossil assemblages that living creatures suddenly appeared representing most all, if not all, phyla to ever exist without a single precursor fossil.

This sudden appearance of multicellular life fosters a problem of Herculean proportions for the evolutionist because all major body types to ever exist appeared out of nowhere without a trace of transitional forms. The late Dr. Stephen Gould, probably the most vocal evolutionist in the latter half of the twentieth century, states the problem and then asks an incredibly salient question:

> Studies that began in the early 1950s and continue at an accelerating pace today have revealed an extensive Precambrian fossil record...**but the problem of the Cambrian explosion has not receded,** since our more extensive labor has still failed to identify any creature that might serve as a plausible immediate ancestor for the Cambrian fauna [animal life]...**Where, then, are all the Precambrian ancestors—or, if they didn't exist in recognizable form, how did complexity get off to such a fast start?** (Gould, 1986, p. 18)

Drs. Henry and John Morris respond to Dr. Gould's question with the answer, "Try creation, Dr. Gould!" (Morris & Morris, 1996, p. 59)

In 1988, science journalist Roger Lewin, tells us:

> ...the most important evolutionary event during the entire history of the Metazoa [multicellular life], **the Cambrian explosion established virtually all major animal body forms (or phyla)...** Compared with the 30 or so extant, some people estimate that the

> Cambrian explosion may have generated 100. (Lewin, 1988, p. 29; emphasis added)

Lewin doesn't hesitate to say that all major body forms or phyla suddenly come into existence; moreover, Professors Valentine and Erwin explain to us:

> If ever we were to expect to find ancestors to or intermediates between higher taxa, it would be in the rocks of the late Precambrian to Ordovician times, when the bulk of the world's higher taxa evolved. **Yet transitional alliances are unknown or unconfirmed for any of the phyla or classes appearing then.** (Valentine & Erwin, 1987, p. 87; also cited in Morris and Morris, 1996, p. 61; emphasis added)

Drs. Henry and John Morris, suggest that:

> The only defense that evolutionists have tried to offer for the absence of any transition to the Cambrian animals is that the latter were the first ones with "hard parts" to be preserved ("hard parts" refers to body coverings, such as shells or an exoskeleton). That, however, is a poor excuse. None of the Precambrian fossils had hard parts (and yet many of their fossils have been found). Furthermore, many of the larger Cambrian fossils did not have hard parts (e.g., jellyfish), yet their fossils are available in considerable numbers. (Morris & Morris, 1996, p. 60)

The Cambrian explosion remains a huge problem for evolutionists, both for the uniformitarian and the punctuationist. However, so does the discovery of **living** "index fossils."

***NOTE TO THE READER:** There is very little difference between an "index" fossil and a "living" fossil. When a living form of an index fossil is found extant, evolutionists call it a "living" fossil.*

Remember, an "index fossil" is a fossil representing a creature believed to have lived only in the rock system where the fossil has been found. This means, the age of "index fossils" for the Cambrian, e.g., trilobites, could be anywhere from 400 to 500 million years. Therefore, when one of these historical markers shows up alive and well, it causes all sorts of problems for evolutionary proponents.

There is a super interesting historical observation mentioned by Dr. Henry Morris and Dr. John Whitcomb, in their book, *The Genesis Flood:*

> It would not be surprising if even the famous trilobite, perhaps the most important "index fossil" of the earliest period of the Paleozoic, the Cambrian, should turn up one of these days. **A creature very similar has already been found.** (1961, p. 179; emphasis added)

Morris and Whitcomb continue by citing an article that appeared in *Science Digest,* December 1957, Vol. 42, entitled "Living Fossil Resembles Long-Extinct Trilobite:"

> A specimen of a "living fossil," perhaps the most primitive **extant** member of one of the major classes of animals, has recently been added to the collections of the Smithsonian Institution. This is a crustacean that has certain characteristics of the long-extinct trilobites, the earth's dominant animals of a half billion years ago, fossils of which are among the earliest traces of an order of life on this planet...Presumably it is exclusively an inhabitant of the mud bottoms of shallow inshore waters and never comes to the surface or has a free-swimming existence. This may account for the fact that it has remained unknown so long. (1961, p. 179)

Could it be that trilobites are still living somewhere in the earth's oceans? It is very possible, because several hundred other creatures thought to be extinct have been discovered alive.

It was reported in the journal *Nature* in March 1993 that a key "index fossil" for the Ordovician, a geological system dated from 400 to 500 million years, was most probably just found in the Pacific Ocean:

> All paleontologists dream of finding a "living fossil." [Researcher] Noel Dilly, it seems, has done so and an account of the discovery appears in a recent issue of the Journal of Zoology. A trawl from deep water off New Caledonia, half way between Brisbane and Fiji, had brought to light an extant [living] pterobranch [colony forming hemichordate, which is any number of small, wormlike, marine animals] that has an astonishing physical resemblance to graptolites, a group considered to have been extinct since the Carboniferous, 300 million years ago...**graptolites are arguably the most important zone fossils [or index fossils] of the lower Paleozoic.** (Rigby, 1993, p. 209; emphasis added)

Many examples could be presented that demonstrate the abundance of, heretofore, thought-to-be extinct creatures offered by evolutionists to prove their model of origins which have been found extant. It is interesting that Drs. Henry and John Morris report:

> A "zone fossil," or "index fossil," is one that is believed to be identified with a specific "geologic age," that its presence in a rock is generally believed to date the rock. As a matter of fact, the very term [index fossil, or living fossil, because both phrases refer to essentially the same idea] is misleading, since most...living plants and animals also have fossil representatives. **In that sense, they are all living fossils.** (1996, p. 114; emphasis added)

It is certainly strange that evolutionists should call these creatures "living fossils." Could someone please tell me what, indeed, is a "living fossil?" I understand the naturalistic twist they have applied to the term. However, I originally learned that when something fossilized, it became mineralized; that it became a rock! Is it not true that fossils are rocks? I suppose when you are attempting to verify something

The Coelacanth

as slippery as naturalistic evolution, it is necessary from time to time to alter the parameters of your terms; to continuously rewrite your dictionary. In this connection, evolutionary biologist Richard Lewontin admits in his review of Dr. Stephen Gould's, *The Mismeasure of Man:* "...I am, myself, rather more harsh in my view. Scientists, like others, sometimes tell deliberate lies because they believe that **small lies can serve big truths**" (Lewontin, 1981). This reality makes evolutionary dogma extremely dangerous!

Nevertheless, it is a huge "naturalistic" stretch or leap of faith, to call something a "fossil" that is still alive, especially when its existence refutes the evolutionary belief about the history and development of the rock record itself. Therefore, the phrase "living fossil" does not make sense in any context, unless you also believe the evolutionary history of the earth.

This brings to mind an article about "living fossils" that I read by the late Dr. Henry Morris, the Father of the Biblical creationist revival in the twentieth century. He carefully documented the fact that creatures from every geological period have been found alive:

> Evolutionists tend to reserve the title **"living fossil"** for those animals and plants which had been considered extinct until suddenly they turn up living today. Consequently, the vast numbers of living

organisms that were already known to be in the fossil record are generally ignored as examples of living fossils. These even include those organisms supposed to be the most ancient of all (the prokaryotes)...

It is significant, that, "Fossils very similar to **living prokaryotes** are found in rocks about 3500 million years old" (Patterson, 1999, p. 129)...

The most important modern prokaryotes are probably the bacteria and the blue-green algae, **and these certainly should be considered living fossils. They have been found in abundance in 3.4 billion year-old rocks from South America.** One wonders why, if evolution really works, these "primitive" organisms have not changed significantly in over a billion years...Until recently, the phylum of vertebrates had been considered a later arrival in evolutionary history. But not now! **Even the vertebrate phylum now extends in the Cambrian period, especially with the recent discovery of two fish in China...**The insects and other land invertebrates are also a very important group, and these practically all seem to be **living fossils...**Whether bees or ants, cicadas or beetles, termites or cockroaches, the fossils of these and other insects are **always practically identical with (though often larger than) their modern descendants.** The same applies to arachnids (spiders) and myriapods (segmented insects as centipedes and millipedes)...

Space does not allow discussion of modern amphibians (e.g. frogs, toads), reptiles (crocodiles, alligators, and turtles), mammals (bats, squirrels, shrews, opossums, tarsiers, etc.), all of which (and many, many others—all alive today) are **practically identical with their fossil representatives...**

There is no space here to discuss the various ages themselves but, in the young-earth model of geologic history, **all the alleged "ages" were actually different deposits of either the Great Flood or the**

> **residual catastrophes following it.** Thus, it is not surprising that the sedimentary rocks laid down by the Flood contain fossils of most of the creatures still surviving in the present age. (Morris, 2000, November 1; emphasis added)

Probably one of the most famous of these so-called "living fossils," one that is used with great frequency in middle and high school textbooks, is the coelacanth (se·lă·kan·th). The coelacanth is a deep-marine, lobe-finned fish, which evolutionists said became extinct about 60 to 70 million years ago.

It is interesting to read the comments written about this fish that were in a state-sponsored high school biology textbook printed by Scott, Foresmen, and Co., Illinois, copyright dates 1985, 1988:

> Until 1938 biologists thought that all lobe-finned fish had been extinct since the days of dinosaurs. Then, off the coast of South Africa, the remains of a peculiar fish were brought ashore by a commercial fisher. The fish was later identified as a coelacanth...It was a species of lobed-finned fish thought to have been extinct for 60 or 70 million years. Living coelacanths have since been found in deep waters east of Africa. **Finding living coelacanths was as surprising to scientists as the discovery of living dinosaurs would be.** (Selznick, Blazer, McCormack, Newton, Rasmussen, 1985 & 1988, p. 472; emphasis added)

Why, do you suppose, the discovery of a living coelacanth was so surprising to evolutionary biologists? Why did they say it was as surprising as finding a living dinosaur?

Their surprise was principally caused for two reasons: First, no one had ever observed this fish alive. It had only been previously observed as a fossil in late Cretaceous rocks. Because this rock system is viewed by evolutionists to represent the history of Earth between 65 and 140 million years ago, the coelacanth was believed to have become extinct sometime near the Cretaceous/Tertiary boundary or about 60 to 70 million years ago (I have seen the date advertised as late as 66 million years ago).

Second, they were surprised because they believed the rock record to be a historical picture of the development of life. They had already decided the lobe-finned fishes were "primitive" because of their position in the rock record. They also made the assumption that their lobed-fins to be the incipient phase (the in-between phase) for the legs that would later be observed associated with the first amphibians. Moreover, because this fish seemed to have disappeared at about the same time dinosaurs were believed to have disappeared, many evolutionists asserted they both became extinct at about the same time and by the same cause. In short, everything they said about the fish was assumed from their evolutionary worldview, which significantly colored their interpretation of the rock record.

In 1938, an angler caught a coelacanth in the waters off the coast of Africa. It was not until 1952 that another coelacanth was caught. However, since then many have been caught, thus ending the evolutionary story of the extinction of the coelacanth. They refer to these fish now as "living fossils" because they feel they have no other way to speak of them and remain faithful to their naturalistic preconception.

It must be noted that the idea of "extinction" can be exceedingly elusive and very difficult to prove. It requires a trained observer looking in all places at the same time to give verification that a particular creature is actually gone from the face of Earth.

The German scientist, Dr. Joachim Scheven, whose Ph.D. is in entomology and paleontology, has gathered one of the world's largest collections of "living fossils"—if not the largest (see the *Creation* magazine, most all issues 1990 to 1995: also see Dr. Carl Werner, *Living Fossils,* vol. 2, 2008). He has amassed hundreds of plant, animal, and insect fossils that correspond perfectly to their identical counterparts living today. Drs.

Henry and John Morris add, that:

> Other publicized living fossils include the metasequoia dawn redwood tree (previously thought to be extinct since the Miocene epoch 20 million years ago), the tuatara, or beachhead reptile (supposedly extinct since the Cretaceous), the segmented mollusk Neopilina formerly thought to be extinct since the Devonian, (300 million years ago), and the brachiopod shellfish Lingula (presumed extinct for about 400 million years, since the Ordovician). (Morris & Morris, 1996, pp. 113-114)

The Wollemi Pine: Another So-called Living Fossil

I think it is eye-opening when an organism thought to be extinct is discovered alive and well; and, reactively, the evolutionary community immediately refers to them as "living fossils." The fact that any of these so-called "living fossils" can be identified with anything alive today is always great support for the Biblical view of creation and time. Their living presence tells us that no significant change has taken place throughout the period of their supposed extinction. This says rather loudly that either evolution has greatly slowed its processes, or it has stopped altogether; or better yet, it never happened at all.

Probably, there is no greater example of the confusion and dismay caused by evolutionary expectations and predictions than was caused in the country of Australia in 1994. It was in the month of September and it was quite by accident that David Noble, an officer with the New South Wales National Parks and Wildlife Services, was hiking in the Wollemi National Park. While hiking, he ventured into a canyon and observed a group of trees growing there he had never before seen. These special conifers were only known from their appearance in Jurassic rocks and were thought to be 135 to 200 million years in age, contemporaneous with the dinosaurs. The oldest estimate for the pines was set at 200 million years of age and, of course, they had been declared extinct.

Wollemi pine

Immediately, the trees were christened "Wollemi pine" (*Wollemi nobilis*) and are now being displayed as one of the rarest examples of a "living fossil" ever discovered. Professor Carrick Chambers, Director of the Royal Botanic Gardens in Sydney, Australia, indicated that finding these trees was like finding a "living dinosaur."

10

Evidences for a Global Flood: Genesis 7-8

Hardly any Lyellian, Darwinian, or uniformitarian assumption (and they are essentially all the same) has escaped serious peer review and scrutiny over the last 50 to 75 years. Predictions based on this model have been found devoid of observable reality, and it is for this reason that the uniformitarian model of Earth history is losing support among many evolutionists. Nevertheless, all evolutionists continue in their evolutionary worldview, even though the evidence is forcing them closer to the conclusions of the Biblical Flood model. They emphatically deny this assessment, but it is nonetheless true.

Because I think it is important for the reader to understand the nature of this evidence and to know a little about the intellectual combat within the ranks of the evolutionary community on this matter, I am offering a brief historical sketch of the data found in the professional literature. This will permit the reader to have some basic understanding regarding the drastic paradigmatic transition in professional geology over the last 50 or so years.

Dr. Henry M. Morris, addressing this aspect of the history of science made these pithy remarks:

> Uniformitarianism—the maxim that "the present is the key to the past" has been the governing principle in historical geology ever since the days of Hutton and Lyell [early nineteenth century], **serving also as the key factor in the rise of Darwinian evolutionism.** Originating as a reaction to the Biblical catastrophism implied by the global Flood of Genesis, it assumed that present geologic processes acting over vast ages of time are sufficient to explain the development of all geologic features of the earth's sedimentary crust.
>
> Modern geologists are now realizing that this approach does not work, and so they are developing what is called "neo-catastrophism," or "episodicity" [referring to episodic occurrences or geological episodes of local to global catastrophes that they believe have caused the observable deformity or alteration of the earth's crust]. **This system postulates intermittent regional—or even global—catastrophes,** accompanied by mass extinctions, all within the standard framework of billions of years. They are still insistent that Biblical catastrophism must be rejected, but it is more obvious all the time that the hard facts of geology do correlate with one worldwide hydraulic cataclysm in the not-too-distant past. (Morris, 1997, p. 275)

To be sure the reader perceives what Dr. Henry Morris has just said, I want to paraphrase his words. He is telling us that Hutton and Lyell dominated the secular, geological worldview for the last 150 years. However, continued observation and research proved their gradualistic (uniformitarian) explanation of Earth to be faulty. As a result, a new model of Earth Science called "neo-catastrophism" is now being more and more embraced among geologists. Neo-catastrophism (see definition at the close of Chapter 6), is a model that postulates the mechanism causing the features of Earth's crust to be episodic local to global catastrophes, which were independent of each other.

I'll begin this documentation in 1966 (44 years since the first edition of this book), just seven short years after Carl O. Dunbar made his assertion that uniformitarian conclusions were universally accepted by all "intelligent and informed people" (see Chapter 3). Because of space limitations, I will only include one or two articles from each decade; but this will be enough information for the reader to observe the transitional move from uniformitarianism to neo-catastrophism.

NOTE TO THE READER: *Dr. Henry M. Morris presents hundreds of evolutionary quotes in his book, <u>That Their Words May be Used Against Them</u> (1997), all of which exhibit the evolutionary confusion regarding this and other origins issues. This book can be purchased from Creation Truth Foundation.*

James W. Valentine wrote:

> **The doctrine of uniformitarianism has been vigorously disputed in recent years.** A number of writers, although approaching the subject from different directions, have agreed that this doctrine... composed partly of meaninglessness and erroneous components... some have suggested that it be discarded as a formal assumption of geological science...It seems unfortunate that uniformitarianism, a doctrine which has so important place in the history of geology, should continue to be misrepresented in introductory texts and courses by "the present is the key to the past," **a maxim without much credit.** (Valentine, 1966, pp. 50-60; emphasis added)

In 1967, P. E. Gretener, Department of Geology, University of Calgary, said:

> Accepting the principle of the rare event (a catastrophic episode) as a valid concept makes it even more desirable to **retire** the term "uniformitarianism," **if further investigations**

> **should prove that singular events of greater importance have indeed taken place in the past, then the term "uniformitarianism" not only becomes confusing but is outright erroneous.** (Gretener, 1967, p. 2205; emphasis added)

In an article published in the *Journal of Geological Education,* Edgar B. Heylmun stated:

> It is hereby submitted that **most scientists are guilty of an over-zealous interpretation of the doctrine of uniformitarianism.** Many instructors dismiss the possibilities of global catastrophes altogether, whereas others ridicule and scoff at the early ideas [the Genesis Flood Model]. These same instructors will implore their students to think scientifically and to develop the principles of multiple-working hypotheses. **The fact is the doctrine of uniformitarianism is no more "proved" than some of the early ideas of world-wide cataclysms have been disproved.** (Heylmun, 1971, p. 35; emphasis added)

Kenneth J. Hsu (shü) and Judith A. McKenzie wrote:

> Catastrophism is enjoying a renaissance in geology. For the last 180 years, geologists have applied consistently an uniformitarian approach to their studies that has stressed slow gradual changes as defined by Lamarck, Lyell, and Darwin. **Now, many of us are accepting that unusual catastrophic events have occurred repeatedly during the course of Earth's history.** The events were significant, since they caused sudden drastic environmental disturbances as well as mass extinctions. (Hsu & McKenzie, 1986, p.11; Hsu was President of the International Association of Sedimentologists at that time; emphasis added)

Professor Derek V. Ager, Head of the Department of Geology and Oceanography at the University of Swansea (Wales), said:

> [that his]...thesis is that in all branches of geology there has been **a return to ideas of rare violent happenings (catastrophes)** and

> episodicity [that took place at various times over the course of Earth history]. **So the past, as now interpreted by many geologists, is not what it used to be. It has certainly changed a great deal from what I learned about it in those far-off days when I was a student...** I must emphasize that I am concerned with the whole history of the earth and its life and in particular with **the dangerous doctrine of uniformitarianism...** This is not the **old-fashioned catastrophism of Noah's flood** and huge conflagrations. I do not think the bible-oriented fundamentalists are worth honoring with an answer to their nonsense. No scientist could be content with one very ancient reference of doubtful authorship. (Ager, 1993, pp. xii, xvi, xix; emphasis added)

I deliberately chose to finish this section with Dr. Ager's quote because of his evolutionary bias. His comments were the most critical of the Biblical position of them all. He plainly tells us that the philosophy of geology has significantly changed since he was an undergraduate student. I have often wondered how long it would take for someone in the evolutionary community to realize the great disparity that existed between the predictions of historical geology and the overwhelming evidence of catastrophe seen in the rock record.

Dr. Ager wants us to know that the new view of evolutionary geology has nothing to do with the Genesis Flood Model. This is because, and it seems obvious to me, he recognizes the incredible similarity between the two views, and he is fearful that creationists will use his comments as evidence for the Bible position. I think it is remarkable that the genuine evidence demands catastrophe, and evolutionists are now acknowledging that evidence.

THE EXTENSIVE NATURE OF SEDIMENTARY LAYERS

There can be no dependable discussion of dinosaurs, or any other Earth Science topic, without some mention of the Genesis Flood. Hence, I want to rehearse, briefly, some of the more obvious effects of a great Flood as explained in Genesis. I want to do this principally because the Genesis Flood is deliberately overlooked by historical and neo-catastrophic geologists; and it causes their conclusions to be colored by a non-Biblical faith.

There is no justifiable explanation for this behavior from a scientific perspective, especially since the entire crustal surface of earth screams of catastrophe. All of the present features display rapid, massive, devastating signs of terrible, tectonic, hydrological activity. To this reality there is no question or debate.

Principle among these signs is the overwhelming depth and width of the bedded sedimentary layers that seem to cover the entire terrestrial portion of the globe. Dr. Harold Coffin reports that in some basins the sedimentary layers reach depths of 60,000 feet, which is over 11 miles. (Coffin, 1983, p. 69)

In addition, Dr. Coffin tells us that the Franciscan Formation found in central California has been measured to be 50,000 feet in depth; and, he wrote that the actual top and bottom of this formation has yet to be completely evaluated (Coffin, 1983, p. 98). Point is: Most of the sedimentary layers of the earth are extremely deep and extensive.

The observed sedimentary nature of Earth's crust appears to have been formed from rapid deposition of water-borne sediments. Therefore, the extensive coverage of many of these sedimentary formations is very difficult to explain without some kind of a significant flood mechanism. Describing the famous Redwall Limestone seen in the Grand Canyon, Dr. Coffin informs us:

> **Rock strata that have broad geological coverage become a strong argument for a worldwide deluge.** The great cliff-forming bed of limestone called the Redwall that is so prominent in the Grand Canyon walls is not restricted to the Grand Canyon area or even to the Southwestern United States. Equivalent beds extend northward to the **Canadian Rockies** and eastward to

> **Kentucky.** The Empire State Building of New York City is faced with limestone blocks from Indiana that correspond to the Redwall. The Mammoth Cave of Kentucky is cut through Lower Carboniferous (also called Mississippian) limestone. Similar material forms strong cliffs in **Britain** and **France.** In **Asia** it crops out in the Himalayan Mountains and caps the top of Mount Everest. (Coffin, 1983, p. 90; emphasis added)

Thus, the extent of strata making up the Redwall Limestone deposit is remarkable and, obviously, cannot be explained as a simple beachfront deposit.

CATASTROPHIC FORMATION OF CANYONS

Earth's many canyons (e.g. the Grand Canyon, located in northern Arizona; the Bryce Canyon, located in southern Utah; or the Royal Gorge, located near Canyon City, Colorado) all appear to be catastrophically formed and fit very nicely into the predictions of the Genesis Flood. The Grand Canyon, in particular, is one of the world's foremost physical wonders. In the book, *Grand Canyon: Monument to Catastrophe,* edited by Dr. Steven Austin, Ph.D. in Geology from Penn State University, we read that the Grand Canyon:

Steven Austin

> ...stretches 277 miles through northern Arizona. The main portion of the Grand Canyon attains a depth of more than a mile and ranges from 4 to 18 miles in width. The north rim, somewhat higher than the south rim, reaches an altitude of 8,500 feet, while the Colorado River cascades along at about 2,400 [feet in elevation]... Within the chasm is a host of pinnacles and buttes, canyons within canyons, and precipitous gorges and ravines...The Christian can

The Grand Canyon

> appreciate, as no one else, the wonders and beauty of God's creation. We can exclaim with David, "O LORD, how excellent is thy name in all the earth" (Psalms 8:1, 9). He has given us the Grand Canyon and its environs as object lessons of His majestic power. (Austin, 1994, pp. 1-2)

The supremacy of the Grand Canyon causes it to be among the most observed of Earth's great chasms. However, its mystery and wonder, the true cause of its popularity, is still a thing of debate and speculation. One's interpretation of this evidence (the evidence is the many features of the canyon itself) completely depends on the particular presupposition that you bring with you to the canyon. Dr. Austin, indicated, that:

> For more than one hundred years, geologists have attempted, in a very deliberate manner, to explain the erosion of the Grand Canyon by **uniformitarian agents.** The elegant notion that the Colorado River eroded the Grand Canyon slowly, during tens of millions of years, has been demonstrated repeatedly to be at odds with the empirical data. Most geologists familiar with the geology of northern Arizona have abandoned **the antecedent river theory**

> [that the Canyon was slowly formed by water erosion caused by the Colorado River]. The less-rational explanation of Grand Canyon, **erosion by stream capture,** involves an accident of incredible improbability [the erosional mechanism of this theory is said to be a run-away gully]...**The catastrophist concept is that the Grand Canyon was eroded by Flood drainage...**The **breached dam theory** is directly supported by evidence of sedimentary deposits from a lake to the east of the Kaibab Upwarp. Furthermore, mechanisms are known, by which floods are able to erode rapidly, even the most solid rock. Numerous landforms of the Colorado Plateau can be regarded as relics sculptured by catastrophic agents associated with rapid drainage of gigantic lakes in the post-Flood period. Today, these landforms endure, through the modern erosional epoch of much-reduced erosion, as monuments of catastrophe. (Austin, 1994, p. 107; emphasis added)

Thus, the formation of Earth's canyons seems to require some kind of catastrophe not known in modern times. Moreover, it is reasonably certain that canyons, of the magnitude of the Grand Canyon, are not formed slowly and gradually—one sand grain at a time—as formerly predicted by uniformitarians. I think it is interesting that as far back as 1938, a geologist named H. B. Baker observed that "the doctrine of uniformitarianism leads to poverty where riches are to be desired" (Baker, 1938; also cited in Austin, 1994, p. 83).

In fact, Professor Earle Spamer makes no bones about it:

> **The greatest of Grand Canyon's enigmas is the problem of how it was made.** This is the most volatile aspect of Grand Canyon geological studies...[the] Grand Canyon has held tight to her secrets of origin and age. Every approach to this problem has been cloaked in hypothesis, drawing upon the incomplete empirical evidence of stratigraphy, sedimentology, and radiometric dating. (Spamer, 1989, p. 39; also cited in Austin, p. 107; emphasis added)

FOLDED AND UPLIFTED LAYERS

The gargantuan uplifting and folding of expansive rock layers observed in some mountains, and to a lesser extent observed in bluffs along rivers and in road cuts, etc., reveals massive forces and conditions that have occurred in the past with which we are not now familiar. Just how can a rock be bent or folded?

I suggest that these worldwide irregularities declare the colossal power associated with the Flood of Genesis. Additionally, I suggest that these massive layers were not folded or bent after they had consolidated or lithified (become hardened into rock), they were folded and bent while they were still soft, damp, and recently deposited, probably during the general uplifting of the continents just after the Flood.

I suspect all sedimentary deposits of the great Flood were originally laid horizontally. There doesn't seem to be any question about this fact; and the massive uplifting forces which produced the bends and folds must have been associated with the late-Flood, particularly the post-Flood mountain building processes. At this time in Earth/Flood history, the newly deposited layers were still wet, soft, and pliable which permitted them to be easily manipulated by the tremendous uplifting pressures that would have bent and folded them like taffy and then hardened them into rock.

Some of the greatest examples that demonstrate the incredible magnitude of this energy can be observed in the Alps of Europe or the Himalayas of Asia. Dr. Coffin observes that:

> The Alps of Switzerland and adjacent countries and the Himalayan Mountains, perhaps more than any other mountain areas, **originated by dynamic and sometimes almost incredible movements.** Geologists have studied the mountains of Switzerland for years, **but the riddle represented by the fantastically folded and contorted strata,** the lack of relationship of some deposits to any surrounding source, and the general scarcity of fossils has been difficult to solve...

The folded sedimentary layers of the Himalayan mountains

> The Himalayan Mountains are of the same type as the Western Alps, but on a much larger scale. Such breaking up of the earth's crust could definitely have been associated with the Genesis Flood. (Coffin, 1983, pp. 157, 160)

Many such examples of widespread catastrophism could be offered to the reader, because whole books have been written about the unthinkable power and geological work performed by the Flood. These phenomena include polystrate fossils, soft-bodied fossils, the abundance of turbidity deposits (underwater mud flows) on the continents, and the growth of the Mississippi River Delta—to name just a few. The list of these features is multitudinous. However, I must hasten on with this book and only have space for one more example, but it fits the present discussion wonderfully.

Fossil Graveyards

There are many extensive burial sites containing hundreds of fossilized bones, representing all sorts of animals, located all over the world. I want to limit the few examples that I will offer of these extensive fossil graveyards to dinosaur graveyards. Drs. Henry and John Morris affirm:

> ...existence of great fossil "graveyards," **in every geological "age,"** clearly speaks of **massive catastrophism in all of these supposed ages...**More and more evidence of rapid burial of fossils and rapid formation of geologic structures seems to be coming in all the time. Great fossil graveyards of mammoths in Siberia, of amphibians in Texas, of hippopotami in Sicily, of fishes in California, Wyoming, Scotland, and many other places and of dinosaurs on every continent are all well-known. (Morris & Morris, 1996 pp. 263-264; emphasis added)

Regarding the fossil graveyards containing dinosaurs, let me say that they are found in many places of the world including New Mexico, Wyoming, Alberta, Canada, and Belgium. World renowned dinosaurian expert, Dr. Edwin Colbert, said that in Belgium "...it could be seen that the fossil boneyard was evidently one of gigantic proportions, especially notable because of its vertical extension through more than a hundred feet of rock" (Colbert, 1968, p. 58; Colbert was Chairman and Curator of the Department of Paleontology at the American Museum of Natural History in New York City) And about the graveyard in Wyoming, he wrote "...the fossil hunters found a hillside literally covered with large fragments of dinosaur bones...in short, it was a veritable mine of dinosaur bones...the concentration of fossils was remarkable; they were riled in like logs in a jam." (1968, p. 141)

Dr. Colbert said, of the graveyard found in Alberta that there were "Innumerable bones and many fine skeletons of dinosaurs and other associated reptiles...quarried from these badlands, particularly in the 15-mile stretch that is a veritable dinosaurian graveyard." (Colbert, 1965, p. 151)

Of the Ghost Ranch quarry in New Mexico, Dr. Colbert wrote that in 1947 a team from the American Museum of Natural History uncovered an immense deposit of Coelophysis (SEE·luh·FYE·sis) fossils (small theropod dinosaurs). Concerning this extensive find, he tells us:

> A large scallop consequently was cut into the hillside containing the almost horizontal bone bed, in order to expose the top of the layer. As the layer was exposed **it revealed a most remarkable dinosaurian graveyard** in which there were literally scores of skeletons, one on top of another and interlaced with one another. It would appear that some local **catastrophe had overtaken these dinosaurs, so that they all died together and were buried together.** (Colbert, 1968, The Great Dinosaur Hunters; emphasis added)

Dr. Colbert also referred to this occasion in a later book he wrote in 1983. In this publication, he explained the same event:

> Dozens of skeletons representing dinosaurs of all ages, from very young individuals to full adults, were found piled up in an almost inextricable mass, necks and bodies, legs and tails intertwined to form a picture of unusual confusion...here is the record, **it would seem, of some natural catastrophe that overtook a herd of early dinosaurs, killing them and burying them at the site of the disaster** [there is very little breakage or disarticulation of the skeletons, which would be the case had the carcasses been washed down-stream]. **One wonders what might have happened on one day two hundred million years ago...**The Coelophysis quarry, which was worked in 1947 and 1948, and again in recent years, has yielded skeletons of almost all ages, from very young individuals to adults. So that this is one of the most completely documented of the dinosaurs. (Colbert, 1983, pp. 33, 34, 63; emphasis added)

The important point in all this is that the reader ascertains that these massive graveyards are found in every supposed geological period, which strongly suggests that all geological "Periods" experienced severe catastrophe and upheaval, or more accurately, that all "zones of habitat" experienced catastrophic upheaval. The Biblical view of fossils teaches that all plants, animals, and insects found in the geological record lived together and died together, and what we observe in the rock record looks very much like the results of the Genesis Flood.

11

St. Peter and Modern Evolution

The Apostle Peter warns us:

> Knowing this first, that there shall come in the last days scoffers, walking after their own lusts, and saying, **"Where is the promise of His coming? For since the fathers fell asleep, all things continue as they were from the beginning of the creation."** For this they willingly are ignorant of, that by the word of God the heavens were of old and the earth standing out of water and in water: whereby the world that then was, being overflowed with water, perished: But the heavens and the earth, which are now, by the same word are kept in store, reserved unto fire against the day of judgment and perdition of ungodly men. But beloved, be not ignorant of this one thing, that one day is with the Lord as a thousand years, and a thousand years as a day. (II Peter 3:3-8; emphasis added)

A chapter reflecting the prophetic content of II Peter 3:3-8 is mandatory to obtain an understanding concerning the godless ramifications embedded in evolutionary uniformitarianism and punctuationism.

Peter tells us that in the "last days" a formidable opposition against the doctrines of the second coming of Christ, the Creation, and the Genesis Flood model will occur (see v. 3). In all reality, this is St. Peter's way of warning believers of a principle battle strategy that will surface in the War between good and evil.

The phrase "last days" seems to refer to a significant period of time that began immediately after the resurrection of Jesus and sometime before the general outpouring of the Holy Spirit during the Feast of Pentecost (Acts 2). It was during this time that Peter tells us these "scoffers" would come.

However, more significant than the timing of the "last days" is the fact that Peter accurately predicts the coming of a form of evolution founded in uniformitarian geology that will deny all three of these Biblical doctrines. His accuracy is spot on, and what's astounding about this prophecy is that Peter's words were written nearly 2,000 years before Darwin and Lyell were born.

Peter said that "scoffers" (a Greek word also translated "mockers") would arise in the "last days," following their own lusts, or "walking" according to their own "passions." Simply stated, Peter is telling us that there will come "in the last days," those who will live and promote a worldview that is based merely on their godless lusts, their desires, and their cravings. It seems consistent with the passage to say that "their lusts" will be the reflection of the greater society in which they live.

Most believers do not see, nor have they contemplated, the indivisible connection between evolutionary doctrines and the authority of God's Word, particularly as it relates to His sovereignty and justice. Peter definitely reveals the cunning nature of this insidious deception.

To reject the Genesis Flood is to not only disavow the absolute authority of the Creator God, but it also disavows His right to sovereignly judge His own creation. Moreover, to interweave the proofs of this rejection with the demands and assessments of credible "science" is to make this godless notion believable and authentic, while reducing the Biblical position to myth, superstition, and ancient religious foolishness. This has happened to millions of people today.

Peter continues by telling us the nature of their argument. They will say, "For since the fathers fell asleep, **all things continue** as they were **from** the **beginning** of the **creation**." This means, since the patriarchs died, which plainly begins with Adam as the first patriarch (Genesis 5), all Earth processes and process rates will remain the same. And this uniformity, they propose, has been continuous since the beginning of the creation. In effect, Peter tells us these "scoffers" allege that the creation was not completed, or perfected, in six twenty-four days as Genesis 1 says. This is blatant evolutionary, uniformitarian dogma!

The motto or catch phrase of this materialistic idea is "the present is the key to the past." In other words, to understand how the present creation has come about, all you have to do is study the present processes, because this creative system is still in action. Thus, the explanation of the physical world can be ascertained through a careful scrutiny of the present processes. Have you heard that before?

That said, it is of greater significance that these coming "scoffers" will, with the content of their mocking, also repudiate the "promise of His coming." Evolutionism is the greatest threat to Jesus Christ and the Bible of any competing philosophical system in the world today—more than communism and more than Islamism, or any other "ism," because evolution is a direct attack against the person, deity, and authority of Jesus Christ, both as Creator and Savior of the world.

Is this condition not similar, even parallel, to the warning Jesus gave Nicodemus in John 3:12, which states that our belief concerning the history of the earth either supports or repudiates our belief concerning Christ's promise of heaven? It is indeed!

Once a person becomes convinced of the truthfulness of evolution in either of its constructs, i.e. life from non-life, uniformitarianism or punctuationism, he also becomes divisively outspoken against the authority of Genesis. But even more than that, he immediately jettisons the "promise of His second coming" as the viable hope for man.

When Genesis is successfully attacked, as it has been by Darwinism over the past 100 years or so, the outcome is the natural deconstruction of the Bible, especially those doctrines concerning man's origin and destiny. No wonder Carl Sagan thought that man was nothing more than a speck of dust lost in space! When a man dies, that's it, it is all over! If there is no supernatural intelligence attending the origin of Earth, it must have required millions and millions of years of Earth history to develop. The important point here is that none of these conclusions are the result of scientific inquiry.

This is a fundamental point. Uniformitarianism was not discovered as a result of Lyell's studying the rocks and fossils. It was, first, a requirement of Biblical prophecy. God knew of its dastardly origin before its scurrilous drivel had ever percolated in the mind of fallen man. Second, it was produced because Hutton and Lyell despised the Mosaic chronology. The founders of uniformitarianism were seeking any alternative to the Biblical record because, as we have shown, they hated the Mosaic record and its God. Third, it was a necessary battle in the War between good and evil to purge the faithful and identify the unbeliever.

It should not be startling that this same idea permeates modern geological circles today. The doctrine of uniformitarianism, epitomized by the motto "the present is the key to the past," even though somewhat weakened, is still a factor in this modern battle between good and evil. This is naturalism gone to seed! It totally takes God and His Word out of the equation. It completely ignores the possibility that a transcendental, loving, purposeful, Creator God exists, that He

Mount Saint Helens on May 19, 1982

is sovereign, self-existing, and can judge man's sin by using a global Flood, or any other mechanism, and that He is the absolute Owner of His creation.

Furthermore, it ignores the geological potential of catastrophic events. However, when Mount Saint Helens erupted, secular men were forced to acknowledge that a catastrophe could accomplish enormous amounts of geological work in a short period of time. The naturalistic model of Earth has significantly begun changing since then.

When the forces that were associated with Mount Saint Helens are magnified thousands of times, as was probably the case in the Genesis Flood, it is inconceivable the amount of geology that can be done in a year. Therefore, to "willfully ignore" any evidence that suggests the legitimacy of that event is to "willfully ignore" a great witness of the rocks and fossils. Moreover, this ignorance is another meaningful effect of uniformitarian geology.

If you have been taught the "millions of years" scenario all your life, that the Bible is allegory at best, and that Christ is just another religious figure among the religious personalities of the world, and then someone comes along and tells you there was a global flood in the recent past, you will think they are nuts. But, what is sad about this story is that it is repeated millions of times every year; and because it is, there is a deceptive blindness over the minds of men that prevents them from seeing their nose in spite of their face. I'll explain this a little later.

We need to further analyze the famous uniformitarian motto, "the present is the key to the past." For the most part, there is very little active geology going on anywhere in the earth today. Therefore, all we observe, most of the time, are the effects of rain, freezing and thawing, wind, erosion, and chemical weathering, etc. These slight actions of nature will certainly not make a Grand Canyon or fossilize a dinosaur. So, when we study present processes as the mechanism explaining all of the observable features of Earth, we are forced to add time, lots of time, to the equation. From these obscure processes, we must deductively reason backwards in time far enough for them to have had any chance to develop the canyons, rock layers, and fossils we see today. In particular, for organic matter (bone, wood, soft tissue) to become fossils, the biggest hurdles to overcome are scavenging and decay because these are the enemies of fossilization.

However, if those slight processes are energized by the sudden, forceful, and abrupt actions of global volcanism, earthquakes, and tsunamis, then evolution's vast time periods dwindle in the wake of this earth-shaking energy and geological capability. When Mount Saint Helens erupted, it represented only a single, very small, fountain of the "great deep." But this one fountain of the "great deep" deposited hundreds of sedimentary layers in just hours and formed canyons. (One canyon, in particular, is one/fortieth the size of the Grand Canyon, and it was carved in volcanic rock in a single day.) In addition, there is a potential coal-forming peat formation at the bottom of Spirit Lake due to all of the tree bark that was catastrophically hurled there in just hours. Mount Saint Helens was a very small catastrophe, but it was a catastrophe, nonetheless, that caused a lot of geological work in a short period of time. It is a grand exhibit that shows the potential of a catastrophe. It is a microcosmic presentation of the geological potential for the Genesis Flood.

How can anyone "willfully forget" a global catastrophe with the magnitude of the Genesis Flood? I don't know, but they do! And this is the only basis upon which an intellectual person can adopt the uniformitarian idea about Earth's features. Peter tells us:

> For this they **willingly forget** that by the word of the God the heavens were of old and the earth standing out of water and in water, by which the world that then existed perished being **flooded with water.** (II Peter 3:5; emphasis added)

This verse is important because it clearly refers to the pre-Flood world—the world that was utterly destroyed by the Flood. It tells us that the antediluvian (pre-Flood) world "perished" by being "flooded with water."

I think it is revealing that Peter tells us the earth before the Flood was out of water and in water (II Peter 3:5). A little trip down any coastal highway helps you to visualize, to some extent, what the Apostle Peter was saying. The terrestrial portion of the pre-Flood Earth was "out of water," but it had its foundation in the ocean. This simply means the crustal material that forms the dry land is also deeply embedded "in water" (consider the islands of the ocean).

The dry land environs of the pre-Flood world were surrounded by water, probably much as it is today. Only the pre-Flood biosphere, some believed to have had more land surface than water surface (probably, just the reverse of what it is today, but we don't know for sure.)

The choice of words Peter used in verse six is important. The word "flooded" in the NKJV ("overflowed" in KJV) comes from a Greek word *"katakluzo,"* a form of our English word cataclysm. The word is extremely forceful, meaning to inundate, submerge, deluge or overwhelm with water; sounds tsunami-like to me!

Furthermore, the word "perished" is from a Greek word *"apollumi,"* meaning to utterly destroy. There have been many excellent books written on the geological effects of the Genesis Flood (none better than the one written by Drs. Henry M. Morris and John Whitcomb, *The Genesis Flood*). However, we still can't imagine the overwhelming intensities and process rates, the unspeakable power, and the devastating consequences of this event. In fact, without God's redemptive plans for Earth, the Genesis Flood could easily have completely destroyed Earth. The Earth survived the Great Flood by grace alone! (Genesis 6:8)

John Whitcomb

Peter gives us hints at how these "scoffers" make this terrible error in judgment. This, no doubt, was because they reasoned from their own "lusts"—their own God-rejecting worldview (a condition that is with us today). Blinded by their naturalistic philosophy, Peter tells us that these men justified their avoidance of God and His judgment by becoming **"willingly ignorant"** (KJV), or they **"willingly forgot"** (NKJV), or it **"escaped their notice"** (NASV), that the earth is full of the evidences of the Great Flood (II Peter 3:5).

The tragedy of overlooking the Genesis Flood leads to a complete misunderstanding of the present Earth with all of its manifold wonders (this I have said many times). It leads to the denial that our Creator God is a God of exacting justice; but, most of all, it alleviates the best evidence of another impending universal judgment. If it is true that there was a global cataclysm such as the one described in Genesis 7 and 8, this testifies to the fact that there is coming another cosmic judgment. Only this time its mechanism will be an "element-melting" fire (II Peter 3: 7, 8-10).

Peter tells us that the **same word of God** that brought the waters of the Flood is reserving the present cosmos (heavens and earth) for the judgmental fires of the future (II Peter 3:7). Simply stated, the only way of deliverance in Noah's day was being on board the Ark. Today, it is still an Ark that provides deliverance, but this time the Ark is a Person and not a boat—the person of the Lord Jesus Christ! We must be hid away with Christ "in God" (Colossians 3:3). If you are not already in the Ark, it's high time to get on-board!

12

What do Fossils Mean?

Have you ever found a fossil, or a curious rock that you thought might be a fossil, and then wondered how it was formed, or what kind of a creature it was, or how long ago it was fossilized? I am sure many readers have asked those questions.

Gary Parker

Dr. Gary Parker tells us that the word "fossil" means "something dug up," and that the people who study fossils are called "paleontologists," from the prefix "paleo," meaning "ancient" (Parker, 2006, p. 6-7).

Most fossils are found in sedimentary deposits such as limestone, shale, and sandstone which are generally deposited by either water or wind. You can easily see the smaller word "sediment" in the term sedimentary, which helps us understand that sedimentary rock is simply made of sediments of other rock.

Sedimentary rock is one of the three types of rock found in the crust of the earth. It is formed when either igneous rock, metamorphic rock, or already formed sedimentary rock is broken up and their loose particles are reworked and cemented together to form new sedimentary rock. The breaking up of these rocks is the result of several different kinds of erosion which are significantly expedited by floods, earthquakes, or volcanoes.

The requirements for sedimentary rock are much like that of manufactured concrete. They both need a cementing substance that is mixed with appropriate amounts of sand or gravel (sediments); but the most important requirement is water. Without water, the sediments, whether igneous, metamorphic or sedimentary, cannot form into new sedimentary rock anymore than the dry contents in a bag of ready-made cement can form into usable cement. Water is the key ingredient.

Most rocks and fossils are formed when organic materials such as wood, leaves, insects, bones, soft tissue, etc. are buried in a mineral solution, generally consisting of lime ($CaCO_3$), or silica dioxide (SiO_2), and water. These ingredients are often called rock cement; and when they appear in the presence of water, they become the necessary solution for the formation of most rocks and fossils (Parker, 2006, p. 8).

There are many types of fossils; but, for this volume, I will only discuss a few of the most common types. Probably, the most commonly found fossils are those of wood, bone, or shell in which one of the above-mentioned mineral solutions fill the pore spaces causing it to become mineralized, a process also called permineralization (Parker, 2006, pp. 9, 79).

Requirements for Fossilization

Dr. John Morris writes that 95% of all fossils discovered are marine invertebrates (such as clams and brachiopods) and of the remaining 5% (4.75% to be exact), 95% of these fossils are mostly plant and algae fossils. However, of the remaining one fourth of one percent that is left, 95% of these fossils consist of other invertebrates, including such creatures as insects (Morris, 1994, p. 70). This means, by comparison, that fossils of higher order creatures such as dinosaurs, mammals, and man are seldom found in the fossil record. They are found, but not in great numbers.

There are two other special groups of fossils significant to our discussion which must be mentioned. One of these is called **ephemeral fossils,** also called trace fossils or **surface fossils.** In this group are animal tracks (birds, amphibians, and reptiles), raindrops, ripple marks, worm burrows, etc.

Trace fossils, which are located at the surface of any particular stratum (rock layer), should not have remained intact except for the rapid, overwhelming burial. Dr. John Morris writes:

> One way to show that only a short time elapsed between the deposition of one bed and the deposition of an overlying bed is to show that the surface features present on the top surface of the lower bed would not last very long if exposed. Therefore, these features had to be covered rather quickly, before they had a chance to erode or be destroyed (Morris, 1994, p. 94).

It seems rather evident that these very shallow and fragile markings were originally made in soft, recently deposited mud or silt. We observe them all over the world; and, in many cases, it is quite noticeable that they were made by very small creatures, or diminutive physical forces. By this, I mean we find fossilized rain drops! These facts tell us that they would have been easily eroded, leaving absolutely no fossil record of their existence if they had they not been buried in a quick and special way. Dr. Henry Morris, adds, "The only way they could have been preserved is by means of abnormally rapid burial (without concurrent erosion), plus abnormally rapid lithification." (Morris, 1984, p. 324)

An example of a polystrate fossil of a tree trunk.

Another special group of fossils that needs to be mentioned are called **polystrate fossils.** The word "polystrate" is a compound word connecting "poly" (many) with "strate" (strata or layers), and it simply means many layers. These unique fossils can be quite large and can consist of animal or plant material (most often huge sections of tree trunks, upwards of twenty or thirty feet or more in length), which transects several sedimentary layers (strata); thus the name "polystrate." "Polystrate trees which extend through more than one layer in effect tie the entire series of layers together into a short period of time." (John Morris, Ibid, p. 101)

Dr. Henry Morris writes:

> Not infrequently, large fossils of animals and plants—especially tree trunks—extend through several strata, often twenty feet or more in thickness...It is beyond question that this type of fossil must have been buried quickly or it would not have been preserved intact while the strata accumulated around it. (Morris, 1984, p. 324)

It is interesting that among the thousands of trees that were deposited in Spirit Lake at the time of the May 18, 1980, eruption of Mount Saint Helens, many of them are standing erect at the bottom of the lake with their root end down as if they grew in that spot. Several of them are already significantly submerged in layers of sediments that have collected in the lake since 1980, setting up the possibility for thousands of polystrate fossils.

Even the secular scientific world observes these phenomena and offers a similar conclusion. Professor F. M. Broadhurst with the Department of Geology, University of Manchester, tells us:

> In 1959 Broadhurst and Magraw described a fossilized tree, in position of growth, from the Coal Measures [deposits] at Blackrod near Wigan in Lancashire [England]. This tree was preserved as a cast, and the evidence available suggested that the cast was at least 38 feet in height. The original tree must have been surrounded and buried by sediment which was compacted before the bulk of the tree decomposed, so that the cavity vacated by the trunk could be occupied by the new sediment which formed the cast. **This implies a rapid rate of sedimentation around the original tree.** (Broadhurst, 1964, pp. 858-869; emphasis added)

It is, therefore, extremely difficult to explain the many polystrate fossils of the world, especially the long sections of tree trunks that transect dozens to scores of strata, without also considering the possibility for rapid, significant burial. The Flood, being both global in extent and abrupt or sudden in power, seems to be the only suitable mechanism that would provide the rapidity of deposition and global coverage necessary to explain this wonder.

NOTE TO THE READER: *There are several pictorial illustrations included in this book which will aid the reader's understanding concerning the compelling nature of rapid burial.*

HOW MUCH TIME IS NEEDED FOR FOSSILIZATION?

Probably, the most misunderstood fact about fossilization concerns the time needed for organic matter to be transformed into fossil material. Fossil formation does not require millions and millions of years. They can be satisfactorily formed in hours to days if the conditions are optimal (and the same is true for the formation of rocks). Moreover, without some very special conditions necessary for fossil formation, even millions and millions of years cannot make fossils. Thus, time is a variable and the proper conditions are the constant.

Dr. Parker informs us that:

> The vast majority of fossils begin forming when a plant or animal was suddenly trapped under a heavy load of water-borne sediment…**It's not the passing of time but the right conditions that forms fossils,** and those conditions are provided by catastrophic flooding. (Parker, 2006, p. 9; emphasis added)

Since the rebirth of catastrophism among secular geologists and paleontologists, many of them are now saying the same things about the requirements of fossilization as creationists. Knowing that permanent fossilization requires deep, rapid burial to safeguard the organism from scavenging and decay, Dr. Anna Behrensmeyer, Ph.D. in Paleontology from Harvard and a professional paleontologist with the National Museum of Natural History in Washington, DC, emphasizes:

> [It is] **mass mortality and instantaneous death and burial [which] create the optimal initial conditions for fossilization,** it is possible that a significant portion of our fossil record is due to such exceptional events [here she refers to catastrophic events or episodes]… **Once an organism dies…there is usually intense competition among other organisms for the nutrients stored in its body. This combined with physical weathering and the dissolution of hard parts soon leads to destruction unless the remains are quickly buried…**These mechanisms [for catastrophic burial] contrast with the **popular**

Photo Gallery

The pictures contained in this gallery are a few of the many fossils and fossil replicas in the CTF Field Collection, consisting in multiple forms. A careful overview of the fossil material in these pictures will help the reader understand and assimilate many of the facts presented in the content of this book.

In particular, I want the reader to pay special attention to the fact that many of the animal and plant fossils displayed in this section have living counterparts today–something that should not exist based on the predictions of evolution. This display is, of course, not exhaustive, but it is a sample of what we find in the field. Further, it is significant that the reader has the opportunity to view ephemeral or surface fossils such as ripple marks, worm borrows, soft-bodied fossils and animal and insect tracks (even trilobite tracks), again surface fossils should not be available today based on evolution. This kind of evidence, I think, reveals the wonder and nature of the Genesis Flood. Like tornados, the Genesis Flood was terribly catastrophic in its global destruction, but also infinitesimally minute in its details.

We send this special section of photography to the reader with the knowledge that most of you will never have the opportunity to accumulate this variety of fossil evidence and with the hope that seeing this evidence, you will be strengthened in your faith concerning the accuracy of the Holy Scriptures.

This is a cast of a **fossil *Tyrannosaurus rex* tooth**. The *T. rex* possessed 50 to 60 cone-shaped teeth, each possessing serrated edges that were regularly replaced. Stan (the Tyrannosaurus at the Black Hills Institute, Hill City, SD) possessed teeth up to 12 inches in length, including crown and root. The T. rex was first discovered in 1902 by Barnum Brown and named by Henry Fairfield Osborne in 1905. It was not until 1988-1990 that their tiny forearms were discovered. It is now debated whether or not the T. rex was a predator, a scavenger, or a combination of the two. The Bible teaches that all created creatures, including the T. rex, only ate herbs and grasses until sometime after the fall and the curse.

This is a research replica of the ***Castoroides ohioensis***—a giant beaver. This great creature has been found from Florida to Canada and represents a phenomenon in the fossil record called gigantism. This beaver, in adult form, was 6 to 8 feet long and weighed upwards to 350 pounds. It had six inch incisors, probably making short work in felling a tree. Uniquely, however, there are other giant creatures in the fossil record: dragon flies, squirrels, armadillos, crocodiles, buffalos, club mosses, etc. These are just the kinds of data one would expect to find based on the Genesis record concerning the pre-Flood earth.

The left image is a **fossil fern** in gray shale. Ferns are the largest group of extant seedless vascular plants known today. The most common of these is found in the *Order Filicales* that evolutionists believe first appeared in the Carboniferous about 300 million years ago. The uplifted lines, seen in the fossil on the right, are **worm burrows** made in limestone. These trace or surface fossils were found in south central Oklahoma near Lake Murray. Their preservation in the rock would require very special burial.

This is a sample of ephemeral or surface fossils. This is a **track-way of a reptile and a scorpion**. Appearing in sandstone, this unique fossil was recently discovered in Arizona. The large track is reptilian and the smaller tracks are scorpion. Without exceedingly special burial conditions, fossils of this type could never be preserved.

This plate exhibits **jellyfish or medusa fossils** from the Wonewoc Formation near the Wisconsin town of Mosinee. This remarkable and unbelievable fossil find was reported in the Abstracts of the North American Paleontological Convention, May 17, 2001. The fossils were verified as late Cambrian jellyfish (more than 500 million years old) and were interpreted to have been stranded. Fossil discoveries as this are intriguing to creationists because there are no hard parts in jellyfish and evolutionists have claimed that soft tissue fossils are not possible. What a testament to the Genesis Flood!

The predominant fossil appearing in the middle of this specimen is a representative of the largest known marine phylum called the **mollusks**. Also known as cephalopods, this group includes such creatures as the octopus, squid, cuttlefish, snails and clams. This particular creature is commonly known as an ammonite. There were many species of these creatures, some circular and some straight. The only ones known to live today are the nautiloids, specifically the Chambered Nautilus of the Indian Ocean and the South Pacific.

This fossil is the **trilobite** believed by evolutionists to be at least 500 million years old. Considered by evolutionists to be an "index fossil" for the Cambrian period, the trilobite is in the phylum of Arthropods. This particular trilobite appears to be in the *Order Redlichiida*. This phylum contains probably 60% to 70% of all the living and fossil creatures in the earth, including spiders, centipedes, crustaceans, insects and scorpions. It has been estimated that the phylum Arthropoda contains more species than all others put together. Dr. Henry Morris noted that some trilobite-like creatures may have been discovered alive.

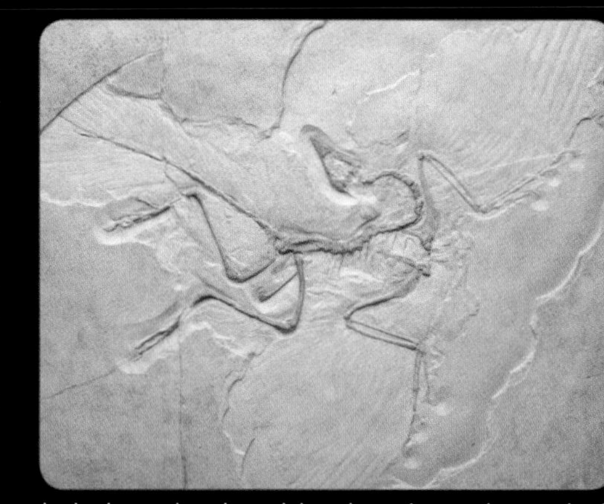

This is a research replica made from the famous 1877 Berlin specimen of the ***Archaeopteryx* fossil** found in the Solenhofen limestone formation in state of Bavaria in southern Germany. It is presented by the evolutionists as the first bird, thus its name meaning ancient wing. However, there have been other bird fossils found in rocks believed to be older than the rocks in which *Archaeopteryx* was found (see http://www.apologeticspress.org/articles/471). Darwin had access to an *Archaeopteryx* fossil just two years after he published the first edition of the *Origin*. It was the London specimen that is now in the British Museum. There have just been seven *Archaeopteryx* fossils found and all of them were found in Germany.

This is a **crayfish fossil** that is advertised to be 100 million years of age. It was found in the Liaoning Province in China. It is generally intriguing that crayfish have been crayfish for a long time.

This is a **fossil crab** in the taxonomic family called *Portunidae* and is said to be Pliocene in age (1.5 to 5 million years). However, there are many extant species of this crab, including the blue crab and the European shore crab. This fossil was found in Rimini, Italy at the Marecchina River.

This is a **Roach** (along with other assorted insects) preserved in Amber that were found in the rain forest of Boyacá Provo, Columbia, South America. Fossil discoveries of this sort are often given ages of 25 million years or more. The assigned ages are spectacular when you observe the similarity of this Roach to modern roaches. It is as if they have not changed in 25 million years. I would suggest they have not changed because the Creator fixed them into genetic patterns that only permits each created kind to bring forth after their own kind.

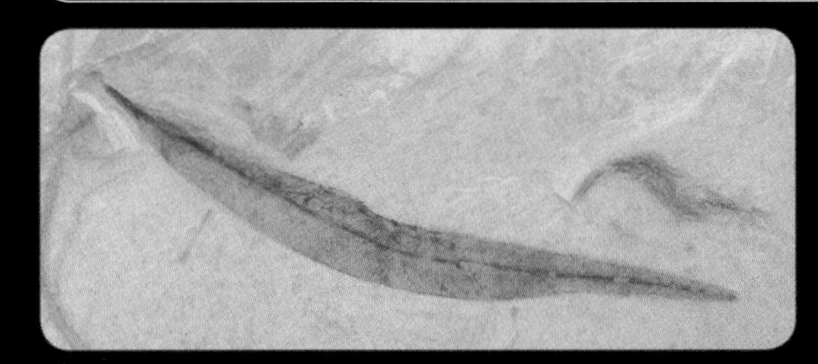

This is a **Green River willow leaf fossil** called a *Salix cockerelli.* It is dated in the Eocene Epoch believed to be 50 million years old.

This **fish fossil** is another sample of the Green river Formation. Of course, all Green River fossils are believed to be Eocene in age (in the vicinity of 50 million years). It is true that most all Green River fossils are well preserved, beautiful fossil specimens. This particular fossil fish is 12 inches in length and is a member of the perch family *(Percidae).* It is a remarkable example of the Green River fossils and a startling reminder (that is, if evolution is true) that perch have been perch for a long time.

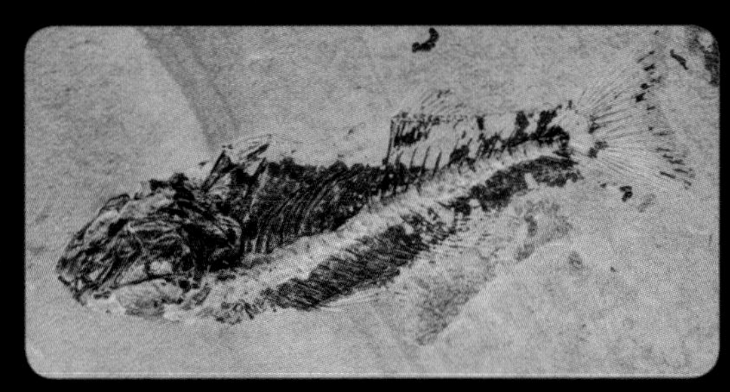

This fossil plate displays a **Green River Formation herring and shrimp fossil**. There is very little difference between the shrimp in this fossil and modern shrimp. Green river fossils are believed to be 50 million years in age. I would suggest the evolutionary assigned dating for these fossils and the mechanism said to have caused them could be in error. The Green River fossils are found in several ancient lake beds that are 7,000 or more feet in elevation. These lakes were probably formed at the conclusion or just following the Noahic deluge during the mountain building processes.

This is a **fossil aspen leaf**. The evolutionary age assigned to this leaf is about 50 million years; however, there are still many species of aspen trees alive and well.

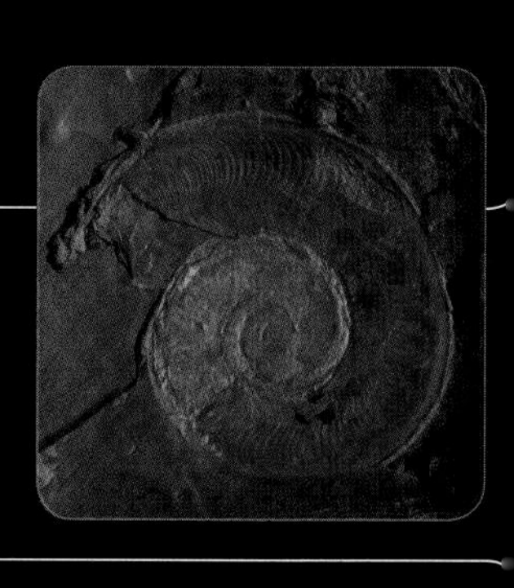

This **Lytoceras Ammonite** is fossilized in black shale from the Holzmaden Black Shale Formation that is located in Stuttgart, Germany.

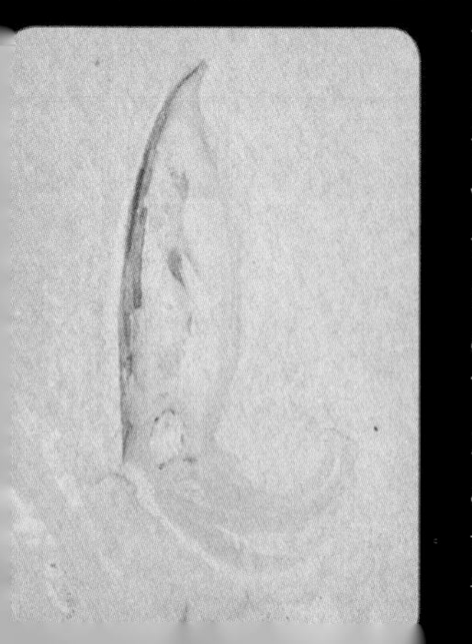

This is a **soft-bodied squid fossil** from Haqel, Lebanon. It was discovered in one of the few famous lagerstätten limestone deposits in the world. The term "lagerstätten" is a German word that refers to place or location. In paleontology, it means a fossiliferous formation that is famous either for diversity of fossils or quality of fossils. The Green River Formation in Wyoming, the Solnhofen Limestone in southern Germany and the Burgess Shale Formation near the little town of Field in British Columbia are among the few that qualify as one of these deposits.

This fossil is a **mammoth tooth** found in Oklahoma. You will notice the distinct difference between the mammoth tooth and the mastodon tooth. The mastodon tooth has distinct raised cusps and the mammoth tooth has a broad, grooved surface. It is believed that this difference reflects the diet of each animal. Both animals were herbivorous; however, the mammoth must have only eaten tender grasses and leaves from small shrubs and the mastodon probably ate a much rougher diet.

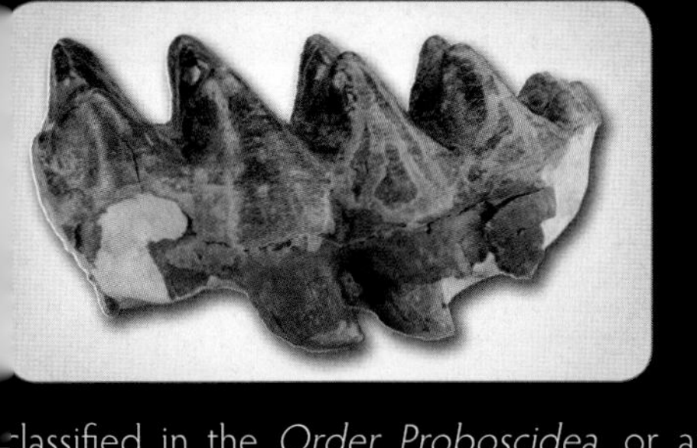

This fossil is a **Mastodon tooth**, the *Mammut americanum* or the American Mastodon. Mastodon fossils are not as plentiful as mammoth fossils, even though both creatures were obviously in the same created kind. All elephant like creatures are classified in the *Order Proboscidea*, or animals with trunks. It seems that the evolutionary bias found its way into the classification of mastodon and mammoths because they have been placed in different families – mammoths are in the *Family Elephantidae* and mastodons are in the *Family Mammutidae*.

This is the result of a small experiment I performed. I was visiting a small town in central France called Clermont Ferrand. I discovered there was a spring in the town with exceeding concentrations of calcium carbonate. I asked the owner of the shop that had been built over the spring if she would help me. She said she would. So, I cut a **small limb from a tree** and took it to her to hang in the water for few weeks. This is the result. She told me she left the limb in the water for six weeks and the portion of the limb that was in the water became completely encrusted with limestone. Fossilization and other related processes are not about time; they are about conditions!

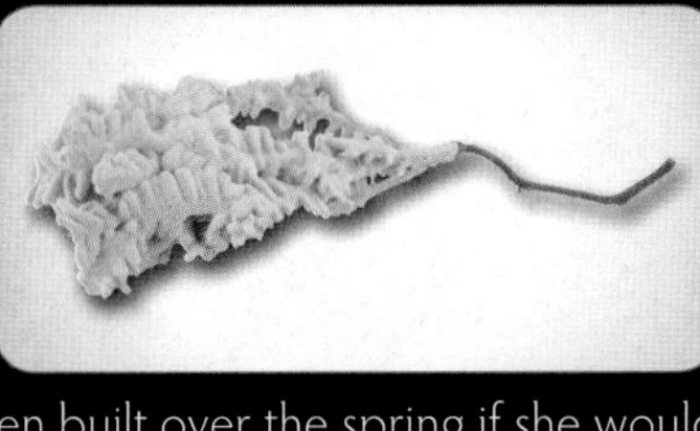

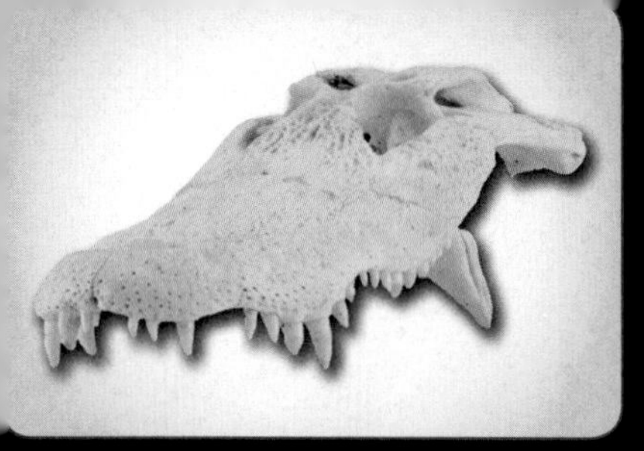

This is not a fossil. It is the **upper skeletal bones of an Alligator skull** from Florida. Alligators represent a variety of reptile in the group called the crocodilians. Crocodilians have been found in the fossil record at the same levels where the dinosaurs are found. Therefore, the evolutionists placed both dinosaurs and crocodiles in the same subclass—*Archosauria*. Since they are both reptilian and they lived at the same time, evolutionary thought places them as close relatives. Have you ever wondered why the dinosaurs are all extinct and the crocs are still thriving?

This is a research replica of the **skull of the short-necked plesiosaur called *Dolichorhynhchops bonneri*** (pronounced Dolly-cō-RING-cŏps bo-NĒ-ī). Discovered in 1976, this marine reptile is believed to have lived in the Cretaceous Period. The complete skeleton is 15.1 feet in length and 13 feet in width and is a part of the CTF field collection. It was found on the South Dakota/Wyoming border near Redbird, Wyoming. The original fossil resides at the University of Kansas in Lawrence.

This picture shows the fossil of the **stem section of *Lepidodendron***. Also called the Scale tree, this fossil stem is claimed to date from the

This **dragon fly** was found in the Crato member of the Santana Formation on the Araripe Plateau in the Ceara State, Brazil. The evolutionary date assigned to this fossil is somewhere between 125 million years to 90 million years of age—in the Cretaceous Period. It is interesting that the fossil dragonfly appears very similar to the modern dragonflies. Wow, dragonflies have always been dragonflies!

This fossil is a section of **dinosaur bone**. Generally, but not always, the main difference between dinosaur bone material and other fossilized bone material is size. The fossilization process is interesting because fossil bones still contain many chemicals and proteins found in fresh bone. This fact alone is a good argument for a young Earth.

This is an **endocranial skull cast of the interior of a *T. rex* skull**. It reflects the actual size of a *Tyrannosaurus rex* brain. It is only about 7 inches in length. The largest section reflects the two olfactory lobes. The average adult *T. rex* skull was about 4 feet in width and 5 feet in length. It has been estimated that its intellectual potential would not be equal to a newly born kitten.

This is a research replica that represents one of several dinosaurs that comprised a larger group called **Dromaeosaurids**. The name means swift or running lizard. Among the different types of dinosaurs that are believed to have populated this group are the *Deinonychus*, the *Adasaurus*, the *Velociraptor*, the *Utahraptor* and the

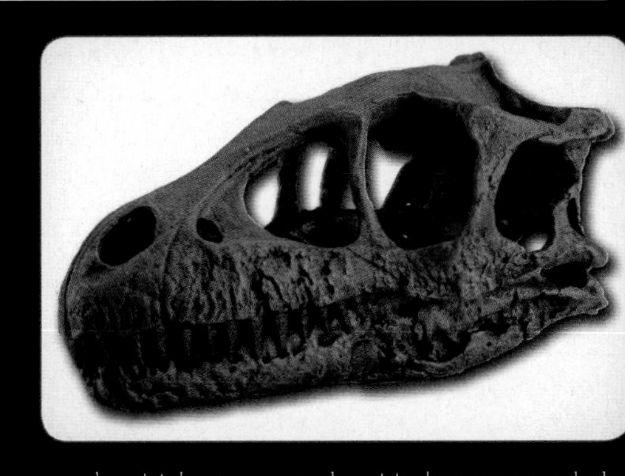

This is another example of a surface or trace fossil known as **ripple marks** that was formed in central Oklahoma sandstone. Ripple mark fossils are probably caused when mud or sand has been exposed to a two-way current flow, as during current or wave action. As the water moved the mud or sand in one direction and deposited it, the returning flow moved and deposited the mud or sand in the opposite direction. This caused the ripple shape. Left undisturbed the cementing content in the sand or mud would then begin hardening and ripple marks are formed.

These fossils are a small portion of a **school of herring** from the Green River Formation. This formation is world renowned for its widespread fossiliferous content that contains plants, vertebrates and invertebrates, even reptiles, mammals and primates. The Green River Formation covers more than 25,000 square miles, reaching from SW Wyoming through Western Colorado and Eastern Utah. The fish in the fossil plate are the very common *Knightia alta*. Evolutionists date this formation from the Eocene time frame.

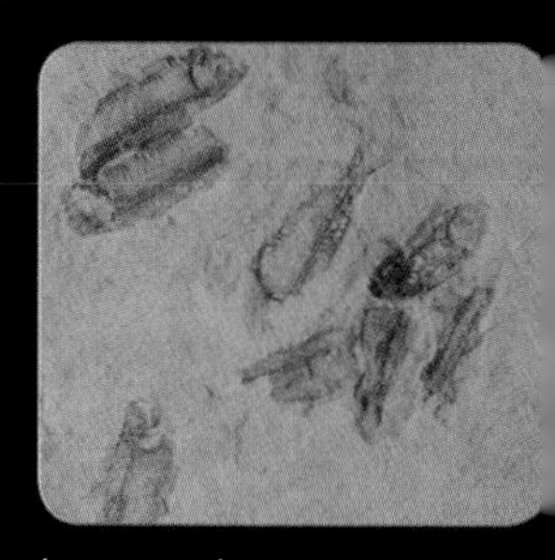

These stones have been identified as Gastroliths, commonly referred to as **stomach or gizzard stones**. Certain animals such as reptiles, birds and some fish that possess gizzards still use sand, grist and small rocks to aid the grinding of food. These particular stones are believed to be Gastroliths from dinosaurs. When the stones became too smooth and polished from the grinding of food, it is believed they were passed through the bowels or regurgitated. Some of the dinosaurs believed to have used gizzard stones are the *Apatosaurus* and the *Seismosaurus* (both sauropods).

This fossil is obviously a **cricket**. Crickets have been classified in the *Family Gryllidae* and have at least 900 known extant species. This fossil is believed to be lower Cretaceous in age (about 100 million years or more). This is one of the oldest cricket fossils known. Therefore, it tells us that crickets have been crickets for a very long time.

This is a **fossil tooth** of the famous, or should I say the infamous, *Spinosaurus*—the pesky old dinosaur that "did-in" the world renowned *Tyrannosaurus rex* in Jurassic Park 3. However, anything said about this dinosaur is very much conjecture because very few fossils bones have been actually discovered. However, thi doesn't bother some evolutionists; they are used to guessing about "facts. Fossil bone fragments for this creature were found in Northern Africa and thi is one of those rare specimens.

This is an exceptional **fossil trackway** – it was made by a trilobite. It was discovered in Mansfield, Indiana in the Mansfield Formation. This is a great example of a trace or surface fossil and would have required very special conditions to have been preserved.

This **fossil bird feather** was found in the Green River Formation at Parachute Creek in Garfield, CO. This spectacular fossil is believed to be Eocene Epoch in age which dates it from 30 to 50 million years ago. There has been much dispute over the origin of flight feathers; however, the fossil record shows feathers fully developed. There is not a trace of incipient stages for flight or feathers.

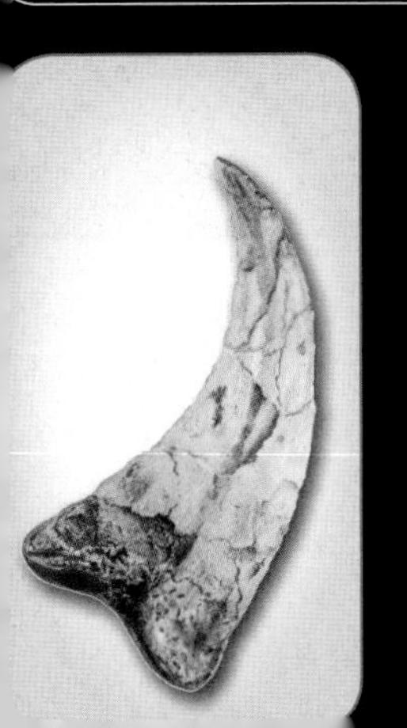

This is a replica of an *Utahraptor* and is advertised to mean "Utah's predator." A **claw core bone** was first discovered in 1991 by Carl Limone. Since then several other bone pieces and fragments have been found by Paleontologists Don Burge and James Kirkland that are believed to be representative of the same animal. It is published that the *Utahraptor* lived 125 million years ago and had feathers, thus the name "raptor" which actually means a bird of prey. This will be convenient "if" feathers are ever confirmed as an anatomical part of dinosaurs. Remember, many evolutionists are convinced that birds evolved from dinosaurs.

The shell to the left is a **pink scallop shell**. It is not a fossil. I found it while walking along the Atlantic coast in 2007. However, the picture immediately right of it is a **fossil scallop** from the same genus *(Pecten)* which contains thousands of fossil species and hundred of living species. The evolutionist believes, because of their interpretation of the rock record, that these bivalves first appeared 230 million years ago sometime during the Triassic Period. However, here again scallops have always been scallops.

This is a significant example of a **Brittle Star fossil** believed to be from the early Jurassic Period about 189 million years ago. Brittle stars have been classified in the *Sub-Class Ophiuroidea* and the true starfish in the sub-class *Asteroidea*. There are hundreds of species of both of these animals living today. Starfish are found in Ordovician limestone. Many evolutionists believe brittle stars and true starfish had a common ancestor; however, as yet this phantom fossil has not been found.

The image on the left is a **shark tooth fossil** from a megalodon shark. It is near 3 inches in length and just less than 2 inches wide. These teeth have been found up to 6 inches in length, indicating that these were very large fish. Some have classified these fish as *Carcharodon megalodon*—the giant shark. Debate continues within the scientific community about the possible relationship between the present Great White Shark and the Megalodon. The image on the right side is a cast of a six inch **megalodon tooth**. One thing is for sure, some time in the past sharks were much larger than they are today.

Few creatures deserve the name "living fossil" any more than the **horseshoe crab**. Found deep in the fossil record, with an evolutionary assigned age as old as 450 million years, this animal is one of the "oldest" creatures for which there are thousands of living counterparts. The fossil in the picture is from the Solnhofen Limestone in southern Germany and has been given a date of 100 million years or more. Their living relatives can be found along the Atlantic coast from Maine to southern Mexico (the American horseshoe crab—*Limulus polyphemus*) and three additional species can be found from Japan to Indonesia.

This is a **fossil leaf** from a Ginkgo tree, sometimes called the Maidenhair Tree, and is another of the world famous "living fossils." Declared to have first appeared in the late Permian Period, more than 200 million years ago, it continues to live on! Indigenous to the orient, it was transported all over the West by the eighteenth century. In fact, the *Ginkgo biloba* (the extant species) can be found growing all over the city where I live in Oklahoma. Remarkable, don't you think?

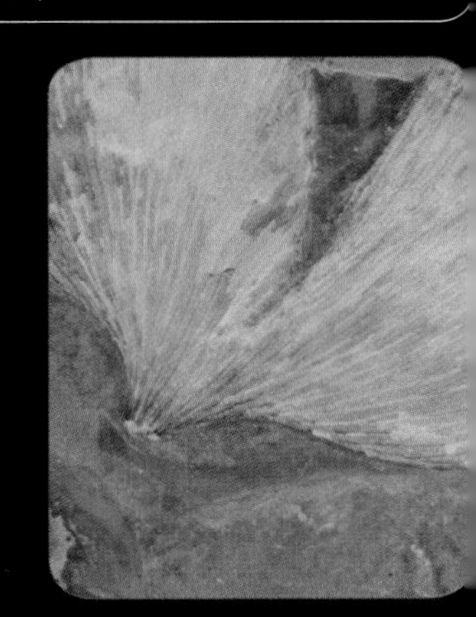

The left picture appears to be a **dinosaur egg** complete with shell, yoke and albumen (this assessment still needs to be verified). However, the right picture is an **Edmontosaurus egg fossil,** which is a member of the duckbilled dinosaur group, also called Hadrosaurs. This fossil egg was found in China

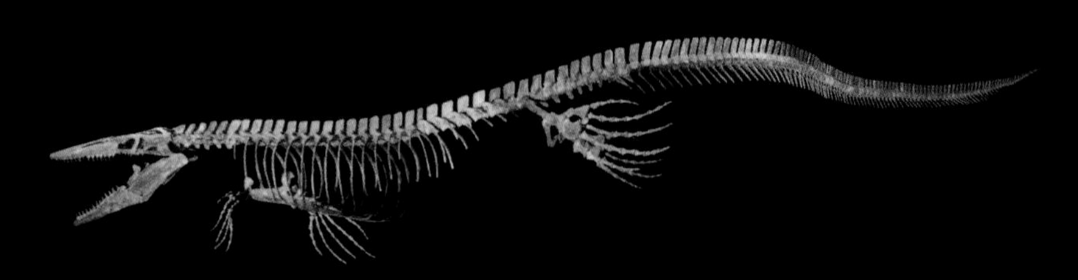

This is a picture of one of the largest Mosasaurs ever unearthed called a **Tylosaur.** Of course, Mosasaurs are marine reptiles and not dinosaurs, nevertheless, they represent one of the large sea creatures created on day five of the creation week. Professor C.D. Bunker from University of Kansas discovered this extremely large specimen in 1911 protruding from the Smoky Hill Chalk in Western Kansas. Field named "Bunker," when the fossil was excavated it proved to be nearly 36 feet in length (12 meters). The picture portrays the present status of this fossil which can be viewed in the Museum at the University of Kansas. CTF has a research replica of the six foot skull of this animal in their field collection. *(Rocky Mountain Dinosaur Resource Center, used with permission)*

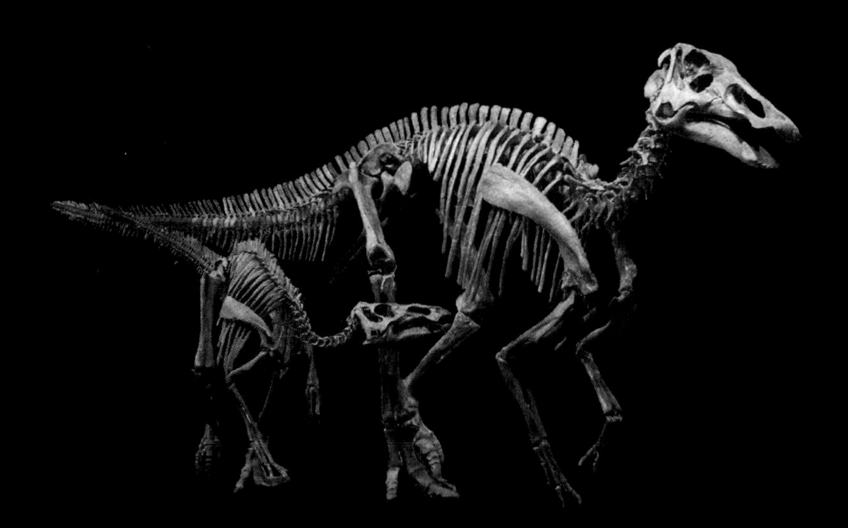

This is a dinosaur called the ***Edmontosaurus***. First unearthed near Edmonton, Alberta, this particular specimen was excavated in Harding County South Dakota. The Edmontosaurus was herbivorous and had four tooth batteries composed of nearly 2,000 teeth. CTF has several different specimens of this creature, including an eleven-foot leg found in Wyoming, a small skull from South Dakota and this 24-foot adult also from South Dakota. *(Rocky Mountain Dinosaur Resource Center, used with permission)*

This skeleton shows the small to medium raptor-like dinosaur called the ***Dromaeosaur***, meaning running lizard. Steven Spielberg made this creature famous in his popular Hollywood movie Jurassic Park. What most do not know is that he portrayed the raptors to be three or more times larger than their real size. These creatures all seem to have the raptor like claw on the inside of their feet. Some evolutionists show these creatures having feathers, but that claim is not verified in fact. CTF has six of these creatures in their field collection.
(Rocky Mountain Dinosaur Resource Center, used with permission)

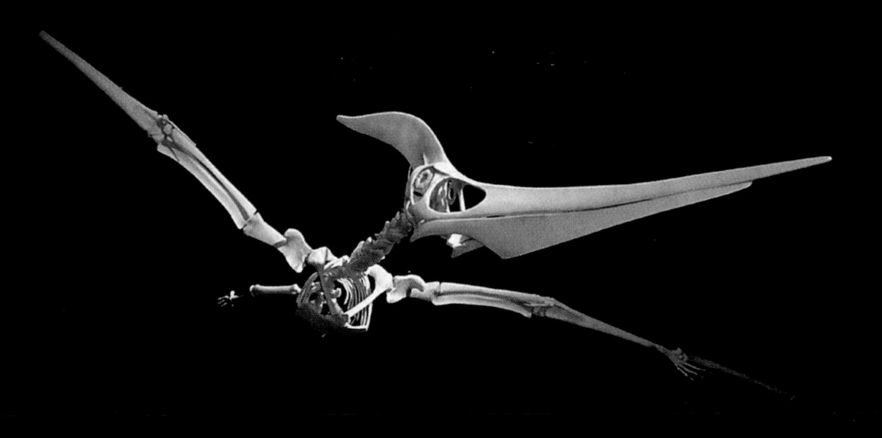

This is a flying reptile called the ***Pteranodon***, meaning winged and toothless. This particular creature was found in the state of Kansas and is twenty-four feet across the wings. It is considered to be male because of the large crest on its head. All flying reptiles are representative of a group of reptiles called Pterosaurs. CTF has two specimens of this creature in their field collection.
(Rocky Mountain Dinosaur Resource Center, used with permission)

This picture shows both the fossil along with a living, fleshed-out replica of a one of the largest bony fishes of the Cretaceous Period called ***Xiphactinus audax*** (meaning sword ray). This creature was found in Western Kansas and was prepared by Triebolt Paleontology. We have a copy of the skull of this specimen in our collection. *(Rocky Mountain Dinosaur Resource Center, used with permission)*

This specimen is another variety of Mosasaur. Smaller than the Tylosaur, this creature is called the ***Platecarpus*** and was also unearthed in Western Kansas. All Mosasaurs seemed to have a double row of teeth located on their palate that has been named the pterygoid that appears to have been used in swallowing their prey. The Platecarpus in the CTF collection is 17-feet in length and is presently on display at the Museum of Earth History in Dallas, TX. *(Rocky Mountain Dinosaur Resource Center, used with permission)*

> **image** of burial as a slow accumulation of sediment through long periods of time, a gentle fallout from air or water that gradually covers organic remains [uniformitarianism]...For remains with simple or complex taphonomic histories [processes of fossilization], **burial is still the most critical step in the process of preservation, and only permanent burial will produce lasting fossils.** (Behrensmeyer, 1984, pp. 558-566; emphasis added)

It is important that the reader carefully note the words of Behrensmeyer. She said that "mass mortality" and "instantaneous death and burial" are the best conditions for forming fossils. Then she said that fossilization has two desperate enemies, the first being **decay** (which she identified as "physical weathering and the dissolution of hard parts") and the second is **scavenging** (that she referred to as "competition for the nutrients stored in the body of the dead animal"). Next, she said the only sure escape is "quick, permanent" burial. This means that fossilization is not necessarily about death. While death is definitely associated with the process, it is really about a unique, catastrophic burial. Sounds like the Flood to me!

I remember visiting a museum that presented a DVD orientation upon entry which informed the guests that the fossilized skeletons we were about to observe were animals that were once alive and are now dead. As I thought about that idea for a minute, it suddenly occurred to me, "Death doesn't cause fossils. Fossilization is caused by sudden, significant burial in the presence of a special mineral solution."

I see many dead rabbits, opossums, and armadillos all along the highways of Oklahoma where I live, but they never make fossils. Of all the buffalos killed on the American frontier, none of them left behind any fossilized skeletons, or parts of skeletons, because fossilization is more about burial and the conditions associated with that burial than it is about death. This is why Dr. Behrensmeyer said that "death" **and** "burial" are the optimal conditions in which fossilization occurs. However, she

A recent discovery of a polystrate fossil near Cookeville, Tennessee.

then asserts that the "popular image" of slow, gradual burial over long periods of time still remains (which is the uniformitarian image still being taught in junior and senior high school textbooks). This is a sad commentary, but it is interesting how very close the secular scientific crowd (Behrensmeyer, et al.) has gravitated towards the predictions of the Genesis Flood Model.

Anomalous Fossils

Anomalous fossils are fossils that are found in the wrong evolutionary order. Fossils are found all over the world, on every continent, at the top of mountains, in the valleys, in rivers and lakes, etc. In fact, I recently had a petroleum geologist tell me that he had just completed some oil exploration and found a clam fossil in the bottom stratum of a core sample that was ten thousand feet down in southern Oklahoma.

Ocean bottom fossils, especially bivalves like clams and brachiopods, are found everywhere. And even though there is some observed and expected order in fossil discovery, as we have indicated, which agrees with predictions of zonal burial (habitat zones), the discovery of **anomalous** fossils (fossils in the wrong evolutionary order) is a predictable expectation in the Flood Model. However, it continues to be an embarrassment to evolutionists. Dr. Henry Morris explains:

> ...anomalous fossils—that is, fossils from one evolutionary "age" found mixed with fossils from another age...is a fairly common phenomenon but is usually ignored or explained away as a case of "reworking" or "displacement," whereby fossils somehow migrate from where they were deposited to where they are found!
>
> This is an ad hoc type of explanation but [it] is hard to disprove and so usually satisfies most geologists since they prefer not to raise questions about the standard geological time scale anyhow... Of course, if the rocks are all of essentially the same age anyhow, as the Bible would indicate...there is no problem [with anomalous fossils]. (Morris, 1984, p. 332)

13

Problems with Extinction

Extinction is a slippery slope because hundreds of thought-to-be "extinct plants, animals and insects" keep showing up alive and well (which is somewhat embarrassing for evolutionists, to say the least). For scientists to know that any particular creature is actually extinct necessitates at least two scientific practices.

The first necessity is a group of specially trained observers, and the second is the requirement that they use modern data-keeping methods. These scientists must carefully scour particular environments looking for evidence of decline and extinction. When it is determined that a particular creature is endangered, this creature is specially identified and carefully protected to prevent extinction. However, when a particular creature is alleged to be extinct that was, heretofore, only known from the fossil record, this is another story.

Extinction is, indeed, very possible; but, it is exceedingly difficult to prove, particularly, for any creature declared to be an "index fossil." To use extinction as an advancement of evolutionary philosophy because the disappearance of certain plants and animals fits nicely into evolutionary expectations about life's history is, to say the least, intellectually dishonest and has proved to be quite embarrassing for evolutionists.

I repeat, this doesn't mean that the extinction of certain creatures is not possible; it is definitely possible for plants and animals to become extinct. It does mean, particularly for the subject of this book, that many of the evolutionary assertions that have served evolutionary purposes have been proven to be inaccurate.

Most of these evolutionary preconceptions find their origin in Darwinian thinking. Darwin believed that if we could be transported back in time far enough, evolutionary processes would have changed the appearance of most creatures until we would not recognize any of the plants and animals we would see. Conversely, he also believed that if we could be transported forward in time, the same phenomenon would occur. He further believed that environmental pressure working with "natural selection" would eliminate the least fit and give rise to another species that was better fit for survival: A process, he believed, that would give rise to myriads of transitional fossils. However, it was this aspect of Darwin's own theory with which he had the greatest difficulty (a fact that we have already covered).

If Darwin's evolutionary expectations proved true, all one should see in the fossil record would be "missing links," or transitional, incipient forms causing total confusion and disorder in the fossil record. Nothing should remotely appear similar in form or feature with animals alive today. But that is not what we see. Most all of the creatures in the fossil record are plants and animals appearing in much the same form with the same features as creatures living today. This fact is indicative that the predictions of Genesis are accurate and true.

There may be varieties within a created kind. For example, in the dog-kind we see poodles, beagles, shepherds, foxes, wolves, etc. because

God placed in each created kind the genetic ability to produce variety within their kind. However, because a kind can only reproduce within its own kind, there can never be transition between kinds. Thus, the Biblical view predicts there can never be transitional forms—there are varieties within kinds, but not transition between kinds. Sound familiar?

There has been considerable confusion caused by evolutionary predictions concerning the supposed extinction of animals for evolutionary purposes. (I have already discussed the problem of "index fossils" and "living fossils.") But does this information also affect the evolutionary prediction of dinosaur extinction? Is it possible we may yet find a dinosaurian reptile alive, even though it seems most of them are extinct?

We have to admit that hundreds of plants and animals, many of which evolutionists considered to be as old or older than dinosaurs that have been declared extinct, have been found alive. All these creatures, according to evolutionary standards, should have disappeared from Earth millions of years ago. Therefore, in answer to my question, I will have to say it is possible that we will yet discover a living dinosaur; however, more research is needed here.

One thing is for sure, the secular scientific world is in great dispute and confusion about the true cause for dinosaur extinction. They even debate whether or not it is possible to know the true cause for their disappearance. I would suggest this dilemma is caused by their "willful ignorance" of the Biblical record of Earth history. Because they have turned their backs on the God of the Bible, they actually don't know what to say about dinosaurs, but they feel they have to say something that sounds "academic and intelligent;" after all, they are the professionals (are they not?).

Are Dinosaurs Extinct? If So, How Did It Happen?

Evolutionists have offered many possible theories to explain the missing dinosaurs. In fact, there are scores of extinction theories available—most of them contradictory—which suggests they really don't know what happened to the dinosaurs (and some evolutionists are willing to admit this)! Scientist/author Glenn L. Jepson probably explains this situation best. In the following article, Jepson lists several of the evolutionary possibilities for dinosaur disappearance:

> Authors with varying competence have suggested that dinosaurs disappeared because the climate deteriorated **[meaning it suddenly, or slowly, became too hot or cold, or too dry or wet], or that the diet did** [meaning there was too much food or not enough of such substances as fern oil; or that there were poisons in water, or plants, or that they ingested harmful minerals; or that beneficial minerals such as calcium or other necessary elements disappeared]. Other writers have put the blame on disease, parasites, wars, anatomical or metabolic disorder [slipped vertebral discs, malfunction or imbalance of hormone and endocrine systems, dwindling brain and consequent stupidity, heat, sterilization, effects of being warm-blooded in the Mesozoic world], racial old age, evolutionary drift into senescent [state of being old], overspecialization, **changes in the pressure or composition of the atmosphere, poison gases, volcanic dust,** excessive oxygen from plants, **meteorites, comets,** gene pool drainage by little mammalian egg-eaters, overkill capacity by predators, **fluctuation of gravitational constants,** development of psychotic suicidal factors, entropy, cosmic radiation, **shift of Earth's rotational poles, floods, continental drift,** extraction of the moon from the Pacific Basin, **drainage of the swamp and lake environments, sunspots, God's will, mountain building,** raids by little green hunters in flying saucers, **lack of standing room in Noah's Ark,** and paleoweltschmerz [a form of sentimental pessimism]. (Lampton, 1989, pp. 117-118; emphasis added)

Pretty amazing, isn't it? And they say the Bible is full of fantasy! However, it is intriguing that many of the above conditions are very consistent with a Global Flood scenario. The preferred belief among these evolutionary ideas, published in most secondary school textbooks for dinosaurian extinction, is either the **Impact Hypothesis** or the results of long-term **continental drift.** Of these two, the majority opinion favors the former notion.

None of these extinction ideas is without serious criticism. In fact, since Jepson wrote the above article, many additional theories have surfaced.

NOTE TO THE READER: *See both editions, 1990 and 2004, David Weishampel, Peter Dodson, and Halszka Osmolska, The Dinosauria, Berkley, CA: University of California Press, 1990, pp. 59-62; 2007, pp. 672-684. Both editions are extremely technical and professional. At the publication of this book, Weishampel was an Associate Professor at The John Hopkins University School of Medicine; Peter Dodson was a Professor of Anatomy at the University of Pennsylvania School of Veterinary Medicine; and Halszka Osmolska was Professor of Paleontology at the Paleobiological Institute in Warsaw.*

Some evolutionists think they have discovered evidence in the fossil record for a 26 million-year-cycle of mass extinction, consisting of many different catastrophic episodes (Weishampel et al, 1990, p. 59). Science journalist Christopher Lampton writes:

> David M. Raup and J. John Sepkoski, Jr., [University of Chicago]... found evidence in the fossil record that there had been many more mass extinctions than had been previously realized...that the number of extinctions went up noticeably every twenty-six million years...What was causing these periodic mass extinctions? There are no known earthly cycles that take twenty-six million years to recur. (Lampton, 1989, pp. 120-121)

Let me remind the reader that all of these evolutionary predictions about mass extinctions are assumptions based on a naturalistic interpretation of the rock record which demands millions and millions of years of Earth history. And, more importantly, they all "deliberately ignore" the Genesis Flood. However, I want to again point out to the reader that most everything they are concluding fits squarely into the Flood Model.

Because of the presence of the element iridium at or near the surface of the Cretaceous period, which many evolutionists believe to be associated with asteroid bombardment, many evolutionists believe these extinction periods are the result of asteroid impact from the heavens. (Lampton, 1989, p.118-121)

NOTE TO THE READER: *The element iridium is a rare element on the Earth that seems to be rather common in meteors. Therefore, it was asserted by some influential evolutionists (particularly Luis and Walter Alvarez) that the iridium associated with the Cretaceous rocks must have come from asteroid impact (Lampton, 1989, p.118).*

However, there is an alternative to the asteroid impact idea regarding iridium enrichment found in the earth. Weishampel indicates:

> An alternative view to extraterrestrial impact, which has gained some support, holds that **major terrestrial volcanism** tapping deep magma sources could be a source of iridium enrichment. (Weishampel et al, 1990, p. 61; emphasis added)

These authors continue:

> Whether a **catastrophe** was generated by extraterrestrial impact or by massive volcanism, the immediate effect must have been **on weather. Either scenario would produce dust clouds in the atmosphere, with attendant darkness and cold stormy weather. When skies cleared and the weather ameliorated a few months or years later, the dinosaurs were gone.** (Weishampel et al, 1990, p. 61; emphasis added)

Are you kidding me? Are they really telling us that a catastrophe—whether asteroid impact or massive volcanism—could have produced major effects consisting of severe weather, rain, and massive dust clouds in Earth's atmosphere that caused winter-like conditions?

These assumptions all agree, in principle, but maybe not in detail, with the Biblical Flood Model. Even though there is much debate about the exact mechanism or mechanisms that lead to dinosaur extinction, the reader must note, as Weishampel has indicated, that sometime in the past there was a monstrous global catastrophe on the earth. I think this is compelling.

Moreover, there is one more wrinkle in this matter that I want to mention. There are some extremely credible scientists and scientific writers that still question anyone's ability to be dogmatic concerning the extinction of dinosaurs. Dr. Edwin H. Colbert, tells us:

> After all of the hauling and backing, after looking at the many theories that have been advanced to account for the disappearance of the dinosaurs, we have made a full circle and are back at the starting point. No one theory and not even a combination of theories can satisfactorily account for the extinction of the dinosaurs, but not the extinction of crocodiles and other reptiles, **for the extinction of late Cretaceous marine reptiles and much planktonic life, but not the ever-burgeoning bony fishes in the seas.** The big-bang ideas of extraterrestrial catastrophes have too many loopholes to provide satisfactory explanations for the disappearance of the dinosaurs. The less spectacular earthbound theories likewise fail to explain the phenomenon in all aspects. In short, the extinction of dinosaurs is a constantly baffling problem, as much of a mystery now as it was a half century ago. **We probably shall never know why these fabulous reptiles, so long the masters of the continents, should have disappeared completely from the earth.** It seems likely that this will remain in the future, as it has in the past, one of the unanswered problems of paleontology. (Colbert, 1983, p. 207; emphasis added)

Dr. Colbert is by no means alone in this assessment. So, what is he saying? Keep in mind that he is an evolutionist and ignores the Genesis Flood. Nevertheless, he is being as honest with the evidence as he can be. He is saying, "How can any catastrophe be selective in its results? Why are the dinosaurs missing and not the animals that lived with them, like the crocodiles, turtles, etc?

In the 2004 revision of *The Dinosauria*—a fabulous and exhaustive work that compiles extensive data about dinosaurs—Weishampel et al. make a similar assessment. After devoting several pages to all the possible scenarios for dinosaur extinction, they conclude with this aloof and dubious statement which I think underscores their own ambivalence about this topic: "We can be sure that whatever else is true, **the last word has not yet been said** vis-á-vis the extinctions at the end of the Cretaceous" (Weishampel et al., 2007, p. 684; emphasis added).

Lampton wrote a very similar idea. After he had thoroughly discussed all known evolutionary models about dinosaur extinction, he admitted:

> **It is possible that we will never know the true reason for the dinosaur extinctions,** or for many similar extinctions in the fossil record... it is likely that new theories will arise to explain these mysterious episodes in the earth's distant past. Probably, the extinction was the result of a combination of factors, and so several of the theories that have been advanced to explain the disappearance of the dinosaurs may be true. (Lampton, 1989, p. 126; emphasis added)

Their confusion is due to their not being able to explain how a global catastrophe can be selective in its annihilation. How could some of the animals apparently be totally destroyed and some not? To this dilemma, Lampton adds:

> If the dinosaurs were such advanced organisms, where are they today? Indeed, whether or not the dinosaurs were warm-blooded, **their disappearance is one of the great mysteries of paleontology.**

> The great Cretaceous extinction is a puzzle for which a satisfactory solution has yet to be found...At the very top of the fossil strata representing the Mesozoic era...At least half of the species living on earth suddenly vanish from the record...All of the saurischians and ornithischians [the name given to the lizard-hipped and bird-hipped dinosaurs] died out.
>
> In fact, all of the archosaurs except for the crocodilians became extinct (archosaurs are a large subclass of reptiles that includes crocodiles, pterosaurs and dinosaurs). Ocean life, from microscopic life-forms to the largest seagoing reptiles, was devastated. On the other hand, the mammals survived, as did most land plants. What sort of event would have eliminated life-forms so selectively, yet devastatingly? (Lampton, 1989, p. 116-117)

Even though evolutionists are inclined to endorse some kind of asteroid impact model, Weishampel et al. still question how an asteroid impact could so devastate the dinosaurs but leave so many other life forms practically untouched:

> It seems tenable to attribute the final disappearance of dinosaurs to an asteroid strike. However, given the massive level of survival of so many organisms (i.e., roughly 85% of marine families and 86% of non-marine vertebrates)...**it is appropriate to question how dinosaurs had become so vulnerable.** (Weishampel et al., 1990, p. 61; emphasis added)

What these authors are saying is a mystery of mysteries. Since we find today 85% and 86% (percentages of extant creatures identified through fossil comparison) of all the known marine and terrestrial animals, respectively; and since the dinosaurs all seem to have been destroyed, how can a catastrophe possess the selective destructive ability to leave all these animals alive while killing all of the dinosaurs? Quite simply, they are asking how can an apparently global event cause the extinction of all dinosaurian creatures (small and great, marine and terrestrial) and leave so many other sea and land creatures to survive to this very day?

I don't want to overstate this problem, or bore you with details, but the fact, timing, and cause of dinosaur extinction represents a significant conflict between Biblical creationists and evolutionists; so, maybe just one more reference. Dr. Bernard Heuvelmans, Ph.D. in Zoology from the University of Brussels, wrote in the Foreword of Dr. Roy Mackal's book, *A Living Dinosaur:*

> ...two of the favorite questions of science writers is **why have dinosaurs vanished** and **what wiped them off the face of the earth?** A plethora of scientific hypotheses, usually of the most extravagant nature, from constipation to a glut of comets, has been put forth over the years to account for this unquestionable extinction. Not a single hypothesis is in the least satisfactory. No wonder; **none** of the scientists responsible for such lucubrations [studied or pretentious expressions] first asked this very simple question: **how could any inner cause or catastrophic impact wipe out selectively a whole group of very diverse animals—two entire orders of reptiles no less—and leave unscathed some of their close relatives, such as the crocodiles and tortoises,** not to mention many other living beings sharing their various habitats. (Mackal, 1987, pp. xi-xii; emphasis added)

Heuvelmans informs us that Dr. Mackal, a distinguished biochemist, engineer, and biologist who received his Ph.D. from the University of Chicago in 1953, also asks very similar questions:

> (1) **Why** should all dinosaurs be extinct? (2) **Could** not at least one genus survive today? After all, we still have the tuatara of the order Rhynchocephalia [which pre-dates the dinosaurs according to evolutionary reasoning], the coelacanth as a sole survivor of the

> Devonian order Crossopterygia, and the nautilus group of which several species still thrive in tropical waters as representatives of an order dating from the early Paleozoic Era [supposedly 400 or more million years ago]. (Mackal, 1987, p. xii; emphasis added)

As I mentioned above, it may seem to the reader that I am seriously overworking this point. However, this is a deal breaker for the evolutionist. For them to admit that iridium appears in many levels of the rock record, and for them to acknowledge that iridium can be provided by volcanism, and for them to admit that there have been many global to regional catastrophes causing mass extinctions at several levels of the rock record, and for them to admit they really don't know why or how the dinosaurs are extinct, means they simply don't know! But, more than this, it also means that the Flood Model has demonstrable evidence which explains these same phenomena in a satisfying manner, if not better.

This brings me to the next taxing question that I will treat in the following chapter. Could there be any dinosaurs still alive today?

14

Could Dinosaurs Be Alive Today?

The contributing factors leading to dinosaur extinction must have included the many significant changes that the earth endured both during and after the Genesis Flood. In reference to dinosaur extinction, I am convinced their decline must have been strongly influenced by the rapid distribution and growth of human population groups associated with the confusion of languages at the Tower of Babel (Genesis 11) after the Flood. Since human pressure on local environments is considered a primary cause for many of the modern extinctions, it would have also impacted dinosaur viability 500 to 1,000 years post-Flood.

The reader must keep in mind that, according to Genesis, God put on the Ark two of every kind of land animal in which was the breath of life. These animals were distinctly land dwellers and would have been animals created on day six of the creation week. The only day five animals to be placed on the Ark (according to the Bible) were the winged animals and they were put on the Ark in groups of seven (probably seven pairs).

There is no record of any marine animals saved on the Ark. This is probably due to the fact that since the Flood was primarily a water event, there was no need to provide a special for place of safety for marine animals, even though many marine animals were killed and buried by giant mud flows (among other forces) and fossilized.

Included in the day-six animals would have been land-dwelling reptiles. All day-six reptiles were terrestrial beings, and it is interesting that all dinosaurs have a unique pelvic design (either bird-hipped or lizard-hipped) requiring them to be terrestrial. Since day six was the day that God created land-dwelling animals and since there is no Biblical account for any other creation event except the first six days of time, dinosaurs must have been created on the sixth day. And, of course, they were included in the animals God brought to the Ark. The language in Genesis 6 and 7 certainly leads me to believe that all day-six animal kinds were saved in the Ark.

There is a minor quibble offered by "gap theorists" at this point. These are old Earth proponents who are convinced of the scientific accuracy for the evolutionary chronology and desire to reconcile millions of years with Genesis 1. They contend that dinosaurs were created in some pre-Adamic age. This prehistoric age, they contend, existed before Genesis 1:3. But it ended when God sent a judgmental, global flood, often referred to as the Luciferian flood, at the fall of Satan from heaven. They further assert that millions of years passed before the re-creation of the heaven and the earth that is recorded in Genesis 1:3. Thus, the title "Gap Theory" refers to this supposed historical gap between Genesis 1:2 and Genesis 1:3. The evidence for this idea is non-existent! In fact, Hebrew grammar, Biblical harmony, and theological consistency will not permit such a fantastic position.

Both Moses and Paul make this clear. In Exodus 20:11, Moses tells us:

> For in six days the LORD made the heavens and the earth, the sea, and **all** (emphasis added) that is in them, and rested on the seventh day....

Paul tells us in Colossians 1:16:

> For by Him were **all** (emphasis added) things created, that were in heaven, and that are in earth, visible and invisible, whether they be thrones, or dominions, or principalities, or powers: all things were created by Him, and for Him. And He is before all things, and by Him all things consist.

Here, as in other passages, the word "all" is interpreted within the bounds of its Scriptural context. Both in the Exodus and the Colossians narrative, the context is inclusive of the entire creation, including the entire space/mass/time continuum. Every star and planet of the heavens, the sun, moon and Earth, the seas of Earth, and every creature intended to live in Earth's biosphere and in the heavens, whether visible or invisible, was created in this six-day period of time. Therefore, the "all" in these verses must refer to every created creature except, of course, God Himself; and this would include the unique reptiles we now call dinosaurs.

All men and land-dwelling creatures that were **not** put on the Ark must have died in the Genesis Flood. I am often asked, "What happened to all the animals that left the Ark after the Flood?" There is no way to know for sure. But the Bible does indicate that they dispersed throughout Earth and came to dwell in areas of compatible habitat. As regards the dinosaurian reptiles, I want to share with you some interesting evidence.

There have been many reported sightings of dinosaurian-type animals in different parts of the earth over the last one hundred years or so. (It must be noted that these sightings are considered "abnormal" only because of the evolutionary speculation about their chronology and extinction.) However, any intelligent inquirer will question their total extinction, principally because of all the other noted "living fossils" that have turned up.

The evolutionary story states rather positively that dinosaurs suddenly became extinct, or that they finally became extinct (depending on which story you are presently reading). At any rate, as I have stated, they allege this extinction was 60 to 65 million years ago; but, as we have learned, they don't know how or why.

The scholarly work by Weishampel et al. states:

> As far as we know, dinosaur extinction was a worldwide phenomenon. **Direct evidence for this event, however, comes principally from two counties in Montana...**Interpretation of a worldwide phenomenon on the basis of data from a single geographic area is obviously very vulnerable. It is to be hoped that new data from localities on other continents will be forthcoming. (Weishampel, et al., 1990, p. 62; emphasis added)

In their 2004 edition, Weishampel et al., indicates that about 110 Cretaceous boundary sites have been identified on the basis of the "irregular" concentrations of iridium (Weishampel et al., 2004, p. 674). However, it must be remembered that even though iridium is a rare commodity at Earth's surface, it is available from volcanic eruptions; and, therefore, its global presence can be explained as an expected residue of the Genesis Flood.

Evolutionists establish this view of dinosaur extinction based on their special placement and explanation of dinosaurs in the rock column. Is the rock record really a long-term historical snapshot of Earth's history? Did dinosaurs really cease to exist at this point in the rock record, or was this the place reflecting their environment or habitat, the place showing their burial and subsequent fossilization due to the Genesis Flood? It all depends on how and why you interpret the rocks and fossils the way you do.

This particular point or line alleged to demonstrate dinosaur extinction drawn across the so-called geological time scale is called the K/T boundary. The acronym KT represents the use of the German abbreviation for the English term Cretaceous. This is point zero for evolutionists—the location of dinosaur extinction. This is the point at which it is said no dinosaur fossils are found higher in the rocks. This needs further investigation, because there have been parts of dinosaurs found in rocks much younger than the Cretaceous (Weisshampel et al., 1990, p. 60). Nevertheless, this supposed line is identified, as we have said, with the discovery of iridium which they believe to be the result of an extraterrestrial impact of some kind (see previous chapter).

Consider this: Pressed on this point at a meeting of the British Association for the Advancement of Science, Professor E. R. Oxburgh, President of the Association's geology section, said, "All such boundaries are arbitrary, and **can be drawn anywhere.**" (Oxburgh, 1983, p. 606; emphasis added) In other words, according to Oxburgh, this so-called boundary between the Cretaceous and the more recent rocks that has been touted for years as a fail-safe demarcation showing the end of the so-called "Dinosaur Age" is just a line on paper, and it can be drawn anywhere you want to draw it. It is not something you can actually find in the rocks; and, as I have shown, since iridium is also associated with volcanic eruptions, its presence can also be an indicator of widespread volcanism or the "breaking up of the great deep" at the beginning of the Flood.

It is important for the reader to understand that the evolutionist cannot allow dinosaurs and man to live at the same time. It is also very important for them to reject anything Biblical; therefore, they pursue an interpretation concerning dinosaurs that "deliberately ignores" the Genesis Flood. If, as the Biblical model predicts, most fossil species represent life as God created them; and if their general location in the rock record represents zonal habitats that existed before the Flood, there should be no problem for dinosaurs and man to have lived at the same time (even though they probably did not occupy the same locations).

If there is Biblical evidence that places dinosaurs with man, and if that evidence can be corroborated with observation (John 3:12), then evolution collapses with its millions of years of Earth history. Furthermore, if one of these dinosaurian-type animals is eventually found alive, and one may be found someday, evolution as taught by Darwin and his colleagues is dead. They know this, and they are aggressively determined in their defense. (In my opinion the presence of the vast array of "living fossils" has already killed any serious support for the evolutionary story.)

An international organization was developed in the latter half of the twentieth century (1982) called the International Society of Cryptozoology (ISC). The research and findings of this organization, even though it is manned by very reputable scientists, has been shunned, even blackballed, by mainstream evolutionary scientists (what Ben Stein calls "Expelled"). Cryptozoology is not generally considered scientifically reliable.

Nevertheless, the sole purpose of this organization is to investigate, analyze, publish, and discuss sightings and other evidence of animals, previously thought to be extinct, rare, or unknown—including dinosaurs. The late president of this organization, who is also a veteran biologist from the University of Chicago, Dr. Roy P. Mackal, shares his experience with one such famous animal known to most of us as the "Loch Ness Monster." Dr. Mackal reported:

> I had become interested in tracking down reports of unknown or unidentified animals in 1965. However, I hadn't been sufficiently impressed to take the matter seriously. During my ten year tenure as one of the directors of the Loch Ness Phenomena Investigation Bureau, which included many summers spent in the field at Loch Ness, Scotland, I became much more receptive toward such data. The Loch Ness Phenomena Investigation Bureau was founded by David James, a member of the British Parliament. The objective of the Bureau was to investigate and, eventually to establish whether or

not there were Loch Ness monsters. After I accepted the invitation to become a director in 1965, I vacillated in my attitude toward the whole matter of reports of unidentified animals and the question of the Loch Ness Monster, in particular. This ambivalent, on-again, off-again mental state I was experiencing vanished forever in September 1970.

On a clear, sunny day, late in the afternoon, Robert Love, Jeff Blonder and I were out in Urquhart Bay on Loch Ness. We were retrieving hydrophones that we had deployed at depths of 100 meters (300 feet) and 200 meters (600 feet), in hopes of recording animate underwater sounds, including any that might be attributable to strange, unidentified animals. Sitting, half-dozing in the bow of Fussy Hen, our work boat, I noted to my right, about 10 meters (30 feet) distant, an upwelling of water, often observed when large aquatic animals approach the surface from below. Instantly, I was fully awake, staring fascinated at what was happening. Without speaking, I pointed in the direction of the disturbance. Robert and Jeff stopped, as though frozen in time as they stared at the boiling water. My pulse raced, my heart pounded—then a black, roughly triangular object, its tip up about 30 centimeters (1 foot) swung out of the water and back down with a splash. About 0.5 meters (1.5 feet) to the right, the convex back of some creature broke the surface, rolling and twisting, lifting the left flipper or fin out of the water as it rotated toward the right. The appendage and the part of the body showing were black to dark brown, rubbery looking, hairless with an occasional wrinkle or roughness. My first thought: "If this is a salmon, it is a mighty salmon indeed—the grandfather of all salmon!" The visible part of the creature that broke the surface ranged in length from 1 to 2 meters (3 to 6 feet). After about a minute, it submerged as quickly as it had appeared.

Robert, Jeff, and I sat stunned, staring at each other, speechless. I recall how for days I refused to admit that I had seen, with my own

> eyes, one of the strange animals of Loch Ness, popularly known as the Loch Ness Monster. Yet this conclusion was inescapable, and, later, when the Academy of Applied Science obtained underwater photographs of these animals, including the appendages, I realized that indeed the triangular object we had observed was the left front flipper, exactly as photographed, and the three of us had not experienced a collective hallucination on that incredible September day in 1970. **From then on, I knew the Loch Ness monsters existed.** (Mackal, 1987, p. 17-18; emphasis added)

The reader needs to take note that Mackal did not attempt to identify the animal because he did not see enough of it. Mackal was an experienced scientist. However, he did say that from that time onward he no longer denied the existence of large animals in the Loch.

Many secular scientists have rejected this new field of science and the views of its proponents for obvious reasons—not the least of which is the fact that many kooks are found in these kinds of environs. Howbeit, I repeat, it is definitely plausible that such an animal may yet be found because there have been so many other organisms found alive that were thought by evolutionists to be extinct.

Much circumstantial evidence has been compiled suggesting the possibility of animals living today that could be dinosaurian. You probably haven't heard about this evidence because it is not widely publicized. Consider the following:

1. *The Illustrated London News,* February 9, 1856, page 166, reported that workmen digging a railway tunnel in France disturbed a huge winged creature at Culmont in Haute Marne while blasting rock for the tunnel (located about 125 miles southeast of Paris)... The creature was described as livid black, with a long neck and sharp teeth. It looked like a bag, and its skin was thick and oily. It soon died, but its wingspan was 10 feet 7 inches. A naturalist

immediately recognized it as belonging to the genus *Pterodactylus anas,* matching the exhibits of known pterodactyl fossils on display in the British Museum of Natural History in London (Doolan, 1993, p. 14).

2. In the damp region of the Congo, called the Likouala region, several scientists confirm findings that point to the possible living presence of a small sauropod dinosaur (possibly of the Brachiosaurus Family). The native population from this region calls this creature the *Mokele-mbembe.* A Congolese biologist, Marcellin Agnagna, reported seeing the creature on May 1, 1983, as he and other members of his team were researching near Lake Tele. Dr. Mackal has written his reports of these matters in a book called, *A Living Dinosaur? In Search of Mokele-Mbembe* (1987); also reported in Creation Ex Nihilo Magazine (1993, Sep.-Nov. 15(4), pp.14-15).

3. While presenting a paper to a joint meeting of the Canadian and American Societies of Zoology in Vancouver, Dr. Paul LeBlond, Professor of Oceanography at the University of British Columbia, said that the Cadborosaurus (nicknamed "Caddy") had been seen by too many people to be ignored. There has been at least one authenticated sighting of the creature every year since the 1930s.

LeBlond believes the native peoples of British Columbia were familiar with this creature because there are many known "rock carvings" and "wooden images" of Caddy dating from 200 AD. Those who have seen this creature describe it to be a "...long-necked beast with short pointed front flippers, a horse-like head, distinct eyes, a visible mouth, and either ears or giraffe-like horns. Often, Caddy is described as having hair like a seal with a hairy mane on the neck. It's serpent-like, up to 7 meters (21 feet) in length, and has been described to undulate just beneath the surface of the water."

Furthermore, the *New Scientist,* January 23, 1993 (Park, p. 16) reported that a "...three-meter juvenile (Cadborosaurus) was apparently removed from the stomach of a sperm whale." Dr. LeBlond and his colleague Ed Bousfield (who works with the Royal British Columbia Museum in Victoria) "...have analyzed the sightings looking for clues to its biology and behavior." They are not sure. They candidly admit, "It could be something like a plesiosaur, the long-necked marine reptile that lived at the time of the dinosaurs." They further indicate that they must keep "their minds open" about the type of animal Caddy might be. "He (LeBlond) further suggests that this is an animal which may be related to some of the sea mammals we know, but because of its habits, we haven't caught one yet. We see them occasionally, and one of these days we'll catch one and it will be one of the known but rare animals of the ocean." (Park, 1993, p.16)

Of course, there needs to be much additional research before this dilemma can be solved. Nevertheless, there is compelling support for the possibility that a few dinosaur-like species could still be alive.

15

The Biblical View of Dinosaurs

I have presented the Biblical view of dinosaurs in scores of religious and secular settings all over the United States and in many parts of the world (some of the readers may have attended one of these meetings). As most of the readers are aware, Creation Truth Foundation (CTF), of which I am founder and president, has gathered one of the largest field collections of legitimate dinosaur skeletons, assorted bones, and skulls in North America.

During these presentations, I have answered dozens of questions about these remarkable creatures, but there are five questions that continue to be asked over and over. I have already provided most of the information that treats the bulk of these questions; however, I want to take meticulous care that Biblical solutions to any unresolved matters concerning dinosaurs are included in this book.

First, I need to enumerate the five questions that are constantly occurring:

1. **Did the dinosaurs really exist?** Or are they just make-believe Hollywood models designed on a computer and made from molding clay? In short, do we really find their fossils today?

2. **When did they exist?** What about the advertised evolutionary notion, which tells us that dinosaurs lived 65 to 225 million years ago during the Mesozoic Era, in a supposed period of Earth history called the "Age of Dinosaurs?"

3. **What happened to them?** There doesn't seem to be any of these creatures alive today, so where did they go, and why are they gone? Do scientists know for sure they are all extinct?

4. **Were any of the dinosaurs on the Ark of Noah?** If they were on the Ark, how did they all fit? Aren't most dinosaurs exceptionally large? How did Noah get them to the Ark? When they got off the Ark, where did they go?

5. **Could any dinosaurs still be alive today?**

To begin with, I must tell the reader that I believe the Bible to be the word of God, every jot and tittle ("jots" are the smallest part of any written form and "tittles" refer to diacritical marks or other exceedingly small markings in written form—see Matthew 5:18). For example, there is a vast difference between the letter P and the letter R, which is directly due to the stroke that extends from the R. Compare the words PAT and RAT. They obviously have completely different meanings, but the only difference between them is the little stroke extending diagonally from the R. This exactitude tells me that the Bible is very accurate. Its history of origins is trustworthy; and although it is not a book of science, when it refers to scientific issues, it tells the truth about them. If the reader has a problem with this view, he or she may wish to stop reading now.

However, if the Bible is the word of the Creator God, its content is not only true, it is necessary for man to know that truth.

In order to provide the proper philosophical context in which to answer these five questions, the reader must keep in mind the comments that I made concerning John 3:12. If I can suitably answer all five of these questions within that context while remaining faithful to the historical predictions of the Genesis Creation/Flood Model, then I have remained faithful to the Biblical record of Earth history. But, more than that, this would significantly undergird the fact that all of the promises of Scripture concerning our eternal destiny are also true. In other words, if the Bible's history is true, then the Bible's promises are true, also.

What I want to do in this section is to see if the Bible's predictions concerning these five questions can bear the scrutiny of earthly observation. I know that in most instances, the evidence offered will be circumstantial. However, when we attempt to answer questions about the "unobserved past," that is about the best we can do (whether creationist or evolutionist).

I will be using my Biblical preconception (faith) to explain what I see in the physical world of observation. I think you will find this to be very satisfying and plausible because the answers will always be reasonable with very little modification or extenuation necessary. Before I answer these five questions, I must first discuss several important precursory issues.

What is Circumstantial Evidence?

The interpretation of origins evidence is worldview driven. Thus, the only practical way to answer questions about dinosaurs, or any other issue from the "unobserved past," is from assumptions or predictions based on one of the two mutually exclusive models of origins—creation or evolution. Even the evolutionary crowd agrees with this fact.

Dr. Douglas J. Futuyma said:

> **Creation and evolution, between them, exhaust the possible explanations for the origin of living things.** Organisms either appeared on the earth fully developed or they did not. If they did not, they must have developed from preexisting species by some process of modification. If they did appear in a fully developed state, they must indeed have been created by some omnipotent intelligence, for no natural process could possibly form inanimate molecules into elephants or a redwood tree in one step. If species were created out of nothing in their present form, they will bear with them no evidence of a former history; if they are the result of historical development, any evidence of history is evidence of evolution. (Futuyma, 1982, p. 197; When this quote was written, Dr. Futuyma was an Associate Professor in the Department of Ecology and Evolution at the State University of New York at Stony Brook; emphasis added)

Please understand that I am not attempting to put words in Dr. Futuyma's mouth. He is not saying that creation is right; quite to the contrary. Now, I don't agree with Futuyma's evolutionary conclusion that any "evidence of history is evidence of evolution." I think the Biblical prediction of variety within a kind satisfies his notion of evidence of history, especially since there are no provable "missing links." I do agree, however, with his notion that only two models are available to explain origins issues. Each of these views (or models) is antithetical to the other; therefore, they cannot be reconciled or harmonized. While both schools of thought study the same evidence, they each interpret the evidence from the predictions or assumptions from their own preconception.

The essential difference between the two worldviews is that creation explains "all things" (Colossians 1:16) as being the result of divine and orderly design, requiring only a week to create, and having only a redemptive purpose. Stated another way, the "divine matter" of creation was redemptive and the "divine manner" was miraculous.

The Creator God took six days to create the universe. He didn't have to, but He did; that's what He said! And when the last person is redeemed, for whom He died, He will destroy this material world with an all-consuming fire in less time than it took Him to create it. And He will instantly create a New Heaven and a New Earth in which the righteous will live forever.

Evolution, on the other hand, explains "all things" as the result of random, natural happenings using the mechanisms of accident and death, having no overarching purpose or design, and requiring billions of years of historical development. Since much of the disparity between these views extends into, and concerns, the history of the earth, the origins debate must consider the legitimacy or the illegitimacy of the Genesis Flood.

To reduce this debate to its ultimate point of incongruity, the disparity invariably depends on whether or not the **Bible** is true concerning Earth history or whether the **opinion of secular science is true.** (Certainly, there are many scientists who are Bible believers.) So, to understand which of these models is the more trustworthy, we must decide for ourselves which view we believe.

Evidence offered in this context is necessarily circumstantial evidence. It is observable, but not testable, and its interpretation is completely controlled by one's worldview. There can be no possible reconciliation established between these worldviews, because each premise is diametrically opposed to the other. We are left with a worldview that predicts cosmos to chaos, or chaos to cosmos; more specifically, it's divine design versus materialistic chance.

Both Models are Faith Based

Because neither model is scientifically provable or disprovable in the pure, rigorous application of the scientific method, we can only evaluate the reliability of each model from their consistency in explaining the observed evidence. Modern science requires observation of cause/effect phenomenon or, in this case, eyewitness accounting of living dinosaurs upon which testing can be done and repeated. This is the empirical process at its best, which finally establishes a fund of information called scientific knowledge about a particular aspect of the observable world. However, for our present subject matter, this scientific rigor is not possible because there are no known dinosaurs to observe and test.

Some would argue that it is possible to observe the effect of some unknown cause (the death of dinosaurs); and, through trial and error, they believe they can reconstruct and explain the cause. In this instance and argument, the "cause" to which they refer is the time and cause of dinosaur existence and the "mechanism" of their extinction. We need to think about this for a moment. What do we really observe about dinosaurs?

We discover their fossil remains in sedimentary rocks all over the world, so we know something about their burial and the requirements for fossilization. We can classify them into common groups, even species, if we discover enough of their skeletal material.

But the time of their origin, their skin color and its make-up, any unique behaviors they may have possessed, the sounds they made, their living habits, the cause of their death and possible extinction, the mechanism causing their fossilization, and so forth cannot be truly known scientifically.

All we can really say about these issues is what we believe about them because we have no observable understanding about the details surrounding the events themselves. This is agreed upon by both groups. There would be no debate at all, except that uniquely in the "big" middle of this controversy is the issue of the sovereignty of the Creator God versus the supposed "autonomy" of man.

To this debate, Professor Jack Horner (lead paleontologist for Montana State University), makes a remarkable statement of truth (please understand Horner is not a creationist at all). He told *National Geographic*:

> For 20 years we've done what we call arm waving. We've made hypothesis based on very little evidence. Now we're sitting down, we're saying, **'we've got all these ideas, are they real?'** (Achenbach, 2004, p. 14; emphasis added)

But here is the mind-boggling problem facing bone hunters:

> The dinosaur fossil record is actually rather poor. Intact, articulated, museum-quality skeletons are fairly rare. Fossils fall apart. A bone exposed to the elements may simply explode. In some bone beds there are so many tiny skeletal fragments you'd think the creatures had been dropped from a plane. (Achenbach, 2004, p. 18)

Thus, Joel Achenbach, the author of the above article, thoroughly explained what few understand—that professional researchers don't know as much about dinosaurian animals as we have been lead to believe they know. In a lengthy article (32 pages), published in *National Geographic* (March 2004) Achenbach said that our lack of scientific knowledge about these animals is far greater than our knowledge about them. Achenbach interviewed nearly two dozen different dinosaur experts and, to his surprise, he received many conflicting stories. Thus, he concluded the article, with these words:

> **...what we don't know about dinosaurs is far more than what we know.** This is still a new and evolving science. We've just scratched the surface. (Achenbach, 2004, p.33; emphasis added)

Why is the scientific community so separated in their views? There is simply no well-defined approach to understanding anything about the unobserved past—the supposed evolution of dinosaurs, notwithstanding.

If you were not there at the beginning, when the first cause produced the first effects, all the testing in the world can never determine the exact conditions and circumstances that were present and operating at that time. In a word, Creation was a miracle, and its processes cannot be analyzed by methods of science. Moreover, if you remove the historical possibility of a worldwide, catastrophic Flood, you are left in a conundrum concerning the present features on the surface of the earth, without also adding millions and millions of years. For our present study, then, the Biblical view includes the Creation and the great Flood of Genesis which happened in the past; and both of which have application to questions about dinosaurs, but neither of which are repeatable, nor observable.

Therefore, the creation story and the evolutionary story about dinosaurs cannot be said to be truly scientific. I agree that scientists approach their work from a strict scientific regimen, but all the research in world cannot reveal the cause, the details, and the mechanisms of the "unobserved past." I can look at a fossil until I am blue in the face, but I will never be able to explain, accurately, how and why it became a fossil even though I may know something about the process of fossilization.

Evolutionists, themselves, contend that dinosaurs both evolved and disappeared from the earth millions of years before man came on the scene. So it is without question, then, that neither of these models satisfies the narrow requirements of the modern scientific purview, and to allege that either opinion is scientific is preposterous. Both views are **belief systems** about the past. Both views require faith—either in the God of the Bible, or in the changeable opinions of unbelieving scientists. So...

Which Model Best Fits the Evidence?

Notice I did not title this section, Which Model Is Right, but which is best. This is how we can measure the consistency of either of the two models. We can compare the predictions from each model with what we actually observe in the world today. Thus, the model that best fits the evidence with the least amount of modification and alteration can be said to be the most reasonable view.

Drs. Henry and John Morris explain this fact in the following manner:

> The fact that we cannot test either belief scientifically, however, does not mean that we cannot discuss them scientifically. We can define two scientific models: a creation model and an evolutionary model. Then compare and explain and, possibly even predict scientific data. That way, we can arrive at a decision as to which model is more likely to be true, even though we can never prove it to be true. **Which faith is the more reasonable faith**—faith in a completed creation or faith in an ongoing evolution? If we can correlate and explain the origin of all scientific data in terms of present processes and phenomenon, then evolution is reasonable. If, however, these data cannot be explained in terms of present natural processes, then one is justified in assuming that they require completed supernatural processes of the past, and this would make belief in creation the more reasonable faith. (Morris & Morris, 1996, p. 14; emphasis added)

What, then, are the possible predictions that can be made about dinosaurs from each of these models?

CREATION MODEL

1. Dinosaurs, like all other created organisms, abruptly appear from the very beginning, fully formed within clearly definable limits of their own unique created kind.

2. The fossil record of dinosaurs would reveal distinct kinds appearing within their own unique paleozone or habitat, without transitional forms between kinds. No missing links!

3. Dinosaur fossils, as all other fossilized remains, would be formed by rapid, significant burial within water-borne sediments.

4. Men and dinosaurs have lived together throughout most of the Earth's history.

EVOLUTIONARY MODEL

1. Dinosaurs evolved from some ancestral animal, possibly some type of ancient amphibian. By gradual descent with modification, governed by natural selection (gene-by-gene, organ-by-organ and limb-by-limb), over millions and millions of years, they came into existence in multiple, unique forms.

2. The fossil record shows a progressive series of animal ancestors leading up to the development of dinosaurs. This development began with fish modifying into amphibians that modified into reptiles that modified into dinosaurs. Many transitional forms should appear in the fossil record associated with each modification.

3. Dinosaurs became extinct about 65 million years ago. Their extinction promoted the close of the Cretaceous Period, and it was approximately 60 to 64 million years before the first man arrived on the Earth.

4. There should be no evidence available that men and dinosaurs ever lived together.

HOW DO WE DETERMINE THE BEST MODEL?

Again, please take note that I did not say which is right, but which is best. I will need to spend a little time discussing the feasibility of these predictions for both models; however, I will not consider them in any particular order. Even though I presented these predictions using dinosaurs specifically, I will explain them in general terms. This is because predictions concerning either creation or evolution operate within their own unique construct—being either Biblical or secular.

Abrupt Appearance, Created Kinds and Missing Links!

The Bible teaches the abrupt, completely mature, fully functional creation of all things right from the beginning (for example, please note the words of Jesus Christ in Matthew 19:4 and Mark 10:6). This is true for all six days of the creation week.

The Bible also teaches that all created "kinds" would only reproduce within their own kind; and while there may be varieties within each kind as we have said, there could not be transition between kinds. This is the Biblical limitation governing reproduction and the differentiation of species. Therefore, the creation model does not predict the possibility of discovering transitional forms ("missing links") in the fossil record or any other place, for that matter. Biblical creationists are not surprised that after millions of fossil discoveries, the much sought for Darwinian transition has not been found.

The act of creation, based on the Biblical Model, also means to create "out of nothing." Technically, this is called *creation ex nihilo* (or creation from nothing), which requires that all created entities must have been fully mature right from the start.

It seems that all creation was originated in the mind of God, and was brought into existence by and through His Word in time. Therefore, the Creator God must have designed everything in eternity and spoke or declared it into existence in time (Psalms 33:6-9). This explains the incalculable design seen in the universe, life, and man. In a strict Biblical sense, to think that the God of the Bible would create in any other manner is nonsense.

The Creation Model, then, requires creation with an "Appearance of Age." Dr. Henry Morris explains:

> Adam and Eve were created as a full grown man and woman, fruit trees were created already bearing fruit, and light rays from stars were created already in transit through space. The whole universe was created "full grown," ready to function according to the divine plan and purpose intended for it by its Creator. (Morris, 1984, p. 175)

Is this borne out by observation? Yes, I think it is. (Much evidence can be amassed in support of this idea. However, room will only permit me to write a brief support.) What do we see in the fossil record? We see fully functional organisms, many of which are within the same kind or a variety of the same kind, of creature alive today. I will illustrate this by citing one of the leading evolutionary scientists whose comments actually support this Biblical fact.

As you recall, it was evolutionary paleontologist and biologist, Dr. Stephen J. Gould, who developed the notion of punctuated equilibrium. This was the best scheme the evolutionists could produce based on observations. What they actually discovered in the rocks were fossils demonstrating fully formed specimens; not Darwinian gradualism at all, but distinct species, fully formed—a condition Gould referred to as "abrupt or sudden appearance." Then, he said, once a particular creature appeared, they remained without change until they disappeared from the rock record—a condition he called "stasis." In explanation, Dr. Gould wrote:

> **The extreme rarity of transitional forms in the fossil record persists as the trade secret of paleontology.** The evolutionary trees that adorn our textbooks have data only **at the tips and nodes** of their branches; **the rest is inference,** however reasonable, not the evidence of fossils... **Paleontologists have paid an exorbitant price for Darwin's argument.** We fancy ourselves as the only true students

> of life's history, yet to preserve our favored account of evolution by natural selection, we view our data as so bad that we never see the process we profess to study...The history of most fossil species includes two features particularly inconsistent with [Darwinian] gradualism.
>
> 1. **Stasis.** Most species exhibit no directional change during their tenure on earth. They appear in the fossil record looking much the same as when they disappear; morphological change is usually limited and directionless.
>
> 2. **Sudden Appearance.** In any local area, a species does not arise gradually by the steady transformation of its ancestors; it appears all at once and fully formed. (Gould, 1977, May, p.14; emphasis added)

This is by no means an admission on their part that the Biblical view is correct. However, it does show the feasibility of the Biblical view. Gould, now deceased, remained a devoted evolutionist and Marxist his entire life. I included this quote simply to illustrate that both creationists and evolutionists are beginning to interpret the evidence in the fossil record in a dramatically similar manner, with only one exception—GOD and His word!

Gould then made a very telling statement in this same article which, again, clearly demonstrates the true nature of good and evil that is deeply inherent in the creation/evolution confrontation and is the reason for the absence of the Creator God in their model. Gould, in no uncertain terms, uses his views about species to support his Marxist beliefs:

> If gradualism [the Darwinian prediction for speciation] is more a product of Western thought than a fact of nature, then we should consider **alternative philosophies of change** to enlarge our realm of constraining prejudices. **In the [former] Soviet Union,**

Karl Marx

for example, scientists are trained with a very different philosophy of change—the so-called dialectical laws, reformulated by Engels from Hegel's philosophy. **The dialectical laws are explicitly punctuational...**Eldridge and I were fascinated to learn that most Russian paleontologists support a model very similar to our punctuated equilibria. **The connection cannot be accidental.** (Gould, 1977, May, p.16)

No, the connection is not accidental, Dr. Gould! The reader must understand that Gould and many of his colleagues were then, and remain, avowed Marxists. Two realities are important for the reader to know here.

First, I must note that Karl Marx and Friedrich Engels found in Darwin's idea of biological survival through struggle, the basis in natural history for their own idea of collectivism or class struggle. Historian Jacque Barzun made this comment:

> It is commonplace that Marx felt his own work to be the exact parallel of Darwin's. **He even wished to dedicate a portion of Das Kaptial** [Marx's book that expressed his views about class struggle] **to the author of The Origin of Species...**in keeping with the feelings of the age, both Marx and Darwin made struggle the means of development. (Barzun, 1958, pp. 8, 170; emphasis added)

I think it is intriguing to know that evolution sought legitimacy from the same philosophy that gave birth to modern communism—the thinking of Marx and Engels (see Henry M. Morris, (2000), *The Long War Against God,* pp. 53-92).

So, we see that this new evolutionary model, called "Punctuated Equilibria," is merely a slight modification of Darwin's original thinking and revealingly found its origin in Marxist philosophy and not science. It left evolutionary time intact—not much really changed. This new model suggests that speciation happens in a method similar to Marx's suggested class change, in a series of rapid events so that transitional features showing these changes in the fossil record are not possible, or are highly unlikely.

Punctuationists argue that throughout the greater part of geological history there is very little detectable change in an organism's development. This period they called "equilibrium." Then, in localized, rare and rapid events a particular species suddenly changes structure and features (due to maintained build up). These events are called punctuations, and it is these punctuations which are believed to be the cause of new species.

In other words, protagonists of punctuationism are admitting that the fossil record shows long periods of stasis or "sameness" because, according to the fossil record, species remain as they first appear and then sudden change takes place leaving no transitional path. In short, they observe nothing new in the fossil record. They simply assume evolution to have taken place and explain the "gaps" (what Genesis calls created kinds) to be the results of this supposed mechanism. Thus, they are explaining the fossil record as it really appears with their own evolutionary twist.

This view is the best the evolutionary crowd can produce based on what is actually observed and, of all things, it found its origin and language in Marxian collectivism (social conflict or struggle and its resolution).

Thus, the question must be asked: Do fossils, "the evidence," justify the story being told by evolutionists? One fact is for sure, there is little difference between Gould's "abrupt appearance" and "stasis" and the creationists' "created kinds" and reproduction only in each kind. So, is the fossil record simply a snapshot of fiat creation showing all kinds of plants, animals, and men that were created fully mature, lived at the same time, and which were overwhelmed by the Genesis Flood in a little more than 365 days? I'll let the reader decide which of the two possibilities they believe to be the more feasible.

16

The Five Dinosaur Answers

To begin this section, let's first take a quick look at the history of the present dinosaur mania and how it all got started. In 1822, an English physician and his wife who loved to hunt and study fossils, Gideon and Mary Mantell, discovered a few strange teeth and bones near Oxford, England. After discussing their find with several informed people, it was observed that the teeth and bones were very similar (though much larger) to those of an iguana—a lizard found in Mexico and South America, among other places.

Realizing that they had found an animal that was, heretofore, unknown to the scientific community, Dr. Mantell named his find Iguanodon, which means "iguana tooth." While Dr. Mantell was dealing with the excitement of his new find, other researchers had located the fossil remains of some kind of large reptile that seemed to be carnivorous. It was soon called "megalosaurus," meaning "giant lizard." Dr. Duane Gish wrote:

> About this same time [referring to the time of Mantell's discovery] some fossil bones and teeth of a huge meat-eating, reptile-like

> animal was discovered...**Since then, the world of science has never been quite the same.** (Gish, 1992, p. 23)

Did dinosaurs really exist? Yes, of course, they did; we find their fossils in every continent of the world. The primary problem facing Bible believers as regards this question is the dilemma concerning man and dinosaurs, and important to this dilemma is the fact that the word "dinosaur" is not found in the Bible.

The Biblical view concerning these wondrous creatures, as I have said, indicates that all land animals were created on day six, the same day God created man. Whereas the evolutionary view contends that dinosaurs didn't show up in the history of life until 300 million years after clams and trilobites and became extinct 60-64 million years before man. Thus, the disparity between to the two views is extreme.

So, why does the word "dinosaur" not show up in the Bible? This is a great question; one that I am asked in every seminar and there is a salient explanation for this fact. It is well known that Dr. Richard Owen (1804-1892), a contemporary of Charles Darwin and a leading comparative anatomist of the time who was the first Director of the British Museum of Natural History, had observed many of the large fossil bones that I have just mentioned. He knew they were reptilian, but they were not associated with modern reptiles. Therefore, in 1841, he proposed at a meeting of the British Association for the Advancement of Science that these fossils represented a separate group of reptiles which he believed to be extinct (Colbert, 1983, pp. 16-17). In this proposal, he suggested that these strange fossils be called "dinosaurian" which means "terrible lizard." Please note his words:

> The combination of such characters, some, as the sacral bones, altogether peculiar among Reptiles, others borrowed, as it were, from groups now distinct from each other, and all manifested by creatures far surpassing in size the largest of existing reptiles,

> will, it is presumed, be deemed sufficient ground for establishing a distinct tribe or suborder of Saurian Reptiles [saurian means lizard], **for which I would propose the name of Dinosauria.** (Colbert, 1983, p. 17; emphasis added)

Thus, the word "dinosaur" was birthed! Ideas about these huge reptiles had occupied the imaginations of people for hundreds of years; however, this was the very first time they had ever been called by the name "dinosaur."

Of course, Dr. Owen recognized their reptilian similarities and chose to uniquely name them from two little Greek words. The word *"deinos"* (the **ei** is pronounced as in the word h**ei**ght or dīnos), meaning terrible, and the word *"saurus,"* meaning lizard.

Richard Owen

Dr. Colbert says:

> What was an obscure scientific term, applied to a few incomplete fossils in 1842 has, within the span of a century and a half become a well-recognized...word, widely used by people around the world. (1983, p. 18)

We now understand the history regarding the development of the word "dinosaur," but does that answer why this famous "word" is not in the Bible? I think it does. Let's think about this for a minute.

When were the primary English versions of the Bible translated? The first English version of Scripture, with any historical import, was translated by John Wycliffe in 1380. The next significant English translation was the work of William Tyndale in 1535, and then the Geneva Bible of 1560. But, probably the most popular English version is the King James Version of 1611. The point is: The 1841 invention of the term "dinosaur" was not available to any of these early English translations. The closest was the King James Version and it was translated 230 years before the word "dinosaur" was coined.

However, it is eye-opening that the Hebrew word *"tanniyn"* (pronounced tă·neen), and the intensive form *"tanniym"* (pronounced tă·neem'), appear twenty-eight times in the Old Testament. Twenty-one of those occurrences were translated by the King James scholars as "dragon." The first mention of the word "Tanniym" is in Genesis 1:21, which Strong says can refer to a marine or land monster (*The New Strong's Expanded Dictionary of the Words in the Hebrew Bible,* #8577, p. 300). Therefore, it seems that a possible nuance of this Hebrew word is very similar to the meaning of the modern word "dinosaur." Dr. Henry Morris tells us that:

> Dragons, for example (Hebrew tanniym), are mentioned at least twenty-five times in the Old Testament. In one of these, the word is used synonymously with "leviathan that crooked serpent," being called "the **dragon** that is in the sea" (Isaiah 27:1). Ezekiel 29:3 refers to "the great **dragon** that lieth in the midst of his rivers." On the other hand, the mountains of Edom are said to have been laid "waste for the **dragons** of the wilderness." (Malachi 1:3) Other references likewise indicate that there were dragons of the desert as well as dragons in the waters.
>
> A number of physiological attributes of dragons are also mentioned. **Dragons** made a wailing sound (Micah 1:8), "snuffed up the wind" (Jeremiah 14:6), and apparently had poisonous fangs (Deuteronomy 32:33). Seemingly, some were...small. Aaron's rod, which is said to have become a "serpent", actually became a dragon. The regular Hebrew word for "dragon" [tannin] is used (Exodus 7:10). Tannin is not translated "serpent" in other passages, at least in the King James Version. Moses' rod, on the other hand, had become a snake

> (Hebrew nahash; Exodus 4:3; 7:15), but the rods of Aaron and the Egyptian magicians became **dragons** (tanniym), presumably small dragons.
>
> Many dragons, on the other hand, were great monsters. The very first use of tanniym is in Genesis 1:21, which is also the first reference to God's creation of animal life. "And God created great (dragons), and every living creature that moveth, which the waters brought forth abundantly, after their kind, and every winged fowl after his kind; and God saw that it was good." **The King James Version translates tanniym here as "whales," but most other versions use "sea monsters" or "sea creatures," but the word is actually "dragons," and the emphasis is on "great" dragons.**
>
> **As a matter of fact, if one will simply translate tanniym by "dinosaurs," every one of the more than twenty-five uses of the word becomes perfectly clear and appropriate...**The only problem with such a translation, of course, is that the dinosaurs are supposed to have died out about 70 million years before man evolved, according to the standard evolutionary chronology. (Morris, 1984, pp. 351-353; emphasis added)

Isn't that intriguing? Dr. Morris plainly indicates that the meaning for both words (tanniym and dinosaur) is interchangeably similar. So similar that Old Testament dragons and the modern reptiles called dinosaurs could very well represent the same "kind" of creature.

Moreover, the Disney-like mythology or superstition regarding the word "dragon" did not influence the King James' scholars, again, for obvious reasons. Dragons to them could have easily meant some kind of dinosaurian, because ancient English history seems to be filled with references to them. (See Bill Cooper, *After the Flood,* (1995) West Sussex, England: New Wine Press, pp. 130-145.)

All kinds of stories have been told about dinosaurs in an attempt to explain their presence on the earth. These stories have ranged from ideas that they didn't really exist; that they were merely invented by scientists and museums; to notions that they were the evil creations of Satan. However, the Biblical worldview about dinosaurs is simply that they were created among the rest of the land dwelling animals on day six of the creation week.

THE DINOSAURS AFTER THE FLOOD

The Genesis Flood must have drastically reworked the surface of the earth, and most of this geological transformation was immediate (hours to weeks). Furthermore, the imbalance suffered in the crust of Earth caused by the dynamic forces of all the global hydraulics of the Genesis Flood would have caused severe residual effects as well (on-going, continuous aftershocks, volcanism, etc.).

The after-shock phenomenon and other residual catastrophic affects must have continued for several hundred years after the Flood, causing, in many places, the continual transformation of the surface of the earth (Consider Genesis 10:25). The following are a few of the Scriptures that help describe the cause of these terrible forces:

> **Genesis 7:11** "...**all** the fountains of the great deep were **broken** up and the windows of heaven were opened..."
>
> **Genesis 6:7** "So the Lord said, 'I will **destroy** man whom I have created from the face of the earth, both man, and beast, creeping things and birds of the air, for I am sorry that I have made them.'"
>
> **Genesis 7:19-20** "And the waters **prevailed exceedingly** on the earth, and all the high hills **under the whole heaven** were covered. The water prevailed fifteen cubits upward and the mountains were covered."

> Genesis 7:24 "And the waters **prevailed** on the earth one hundred and fifty days."

Among the unique features of today's Earth that are directly due to the Flood are the continental divisions of the earth, which all scientists agree were together at one time and, formed one large contiguous terrestrial environment. The perfect nature of Earth's original climate was severely changed, including significant temperature zones and weather patterns that did not exist before the Flood (consider Genesis 8:22—this would include rapidly changing thermal fronts causing severe storms producing hurricanes, tornados, highly pronounced winter seasons, etc.). Furthermore, the residual effects of the Flood, which would include continual tectonism, volcanism, and developing glaciations causing the great ice age probably continued for 500 to 1,000 years.

These effects, plus a developing disharmony of the earth's environmental balance, must have caused devastating consequences on all life forms. Moreover, these environmental, climatic and habitat transitions would have affected some life forms more than others. This would have absolutely caused epidemic extinctions of all kinds of aerobic (air-breathing) life—an effect that continues to this very day.

It is believed that animals are becoming extinct daily. Extinction, therefore, has been the fate of many of the animal kinds and particular varieties of kinds, which originally were preserved by Noah's Ark.

The violent nature of the Genesis Flood would predict the kinds of features seen in the rock record today, and may have included asteroid bombardment (many creationists believe it did—all the features mentioned in the last few paragraphs are found in most of the evolutionary explanations for the extinction of dinosaurs). The bottom line seems to be that all dinosaurs, not on the Ark, died from a global, catastrophic overwhelming event of some type, and the Bible identifies this to have been the Genesis Flood.

Once the Flood was over and the animals were safely off the Ark, the dinosaurs, as well as other animals, must have been seriously impacted by human pressure (consider the possible results of the divine promise referred to in Genesis 9:2). Remember, it is human pressure that is blamed for most all animal extinctions today.

You may have some questions at this point:

1. How do we know dinosaurs were put on the Ark?

2. How could enormous animals the size of dinosaurs fit on the Ark?

3. How do we know dinosaurs were on the Ark?

I believe God put dinosaurs on the Ark because they were among the terrestrial reptiles created on the sixth day of creation, and Genesis 6 and 7 tells us that God brought to the Ark two of all creatures in which was the "breath of life." Since God created all land animals on day six, it should not be too difficult to understand this would also include the reptiles we now call dinosaurs: But what about their enormous size?

Small in Size

God would not have brought any fully mature dinosaurs or mammals to the Ark for at least two reasons. First, this behavior would not have been in agreement with His intent. He brought the animals to the Ark to save their seed, in order to replenish Earth with their kind after the Flood.

It is well known today that reptiles, in varying degrees, grow every year they are alive. They only quit growing when they die. Thus, we believe since dinosaurs were reptilian, they probably responded in the same manner.

The excavators of the famous T. rex fossil named Sue that was found in the rock layers of Harding County, South Dakota, explained that her enormous size was due to her old age:

> The pelvis indicates that **Sue lived a long life.** The ischia and pubes (these are two of the bones that make up the pelvic girdle) in most tyrannosaurids are separate, but Sue's are fused. In living animals such bone fusion is most commonly seen in **older** individuals. (Webster, 2000, p. 28; emphasis added)

Plainly, God would not have loaded grandfather and grandmother dinosaurs on the Ark (for that matter, He would not have placed older animals of any kind on the Ark), because this would have worked contrary to His conservation purpose. He wanted them to be prime, fertile animals so that when they disembarked from the Ark they would be at the beginning of their reproductive maturity. This purpose obviously doesn't fit well with grandfathers and grandmothers. (Remember, reptiles today grow every year they are alive and since dinosaurs are considered reptilian they must have possessed the same growth behavior.)

Fully-grown sauropods (long necks), theropods (three-toed carnivores), ornithopods (this group included the duck-billed hadrosaurs), stegosaurids (stegasaurs), and ceratopids (included the Triceratops) were all extremely large as mature adults. This will help you understand my remark about grandfathers and grandmothers because dinosaurs were at their largest size in old age. Therefore, they must have entered the Ark near the beginning of their sexual maturity (much smaller than their full mature size). Remember, once on board the Ark, all animals had several more months to mature.

However, some kinds of dinosaurs did not grow to these enormous sizes no matter how long they lived. For example, the Microceratops only reached a length of 30 inches and the Alocodon 3 feet, etc. God probably saved yearlings or teenagers. To be sure, they were younger animals and, of course, much smaller than is reflected in their fully mature-sized fossils.

Actually, most dinosaurs were small. Based on actual fossil evidence, the average-sized adult dinosaur is about the size of a sheep or a small cow (Sattler, 1983, p. 149). Newly born tridactyl (three-toed) bipeds and newly born sauropods were no larger than modern dogs.

This is interesting. Scientists have found and identified the eggs of a sauropod called Hypselosaurus (pronounced, hip'sel·uh·saw'rus, meaning high lizard). The eggs were found in southern France and were extremely small—only 7 to 12 inches long and 6 to 10 inches wide. The Hypselosaurus could grow to a height of 18 feet, a length of 40 feet and a weight of 20,000 pounds (10 tons). BUT, when born, they weighed a mere 2 pounds (*Dinosaurs,* 1987, pp. 195-197). Wow! Our God is indeed the epitome of wisdom. The Hypselosaur eggs were approximately the dimension of this book you are now reading.

The growth rate of dinosaurs was significantly faster than humans. A human increases in size about 20 to 25 times from infant to adulthood, whereas the Hypselosaur increased in size about 10,000 times during the same period (Dinosaurs, 1987, p. 197). This provides an explanation for how God could have placed these terrible lizards successfully on the Ark of Noah in their juvenile form.

JEREMIAH 33:19-21

It is abundantly evident that the Bible teaches that God created all air-breathing animals and man on the fifth and sixth days of the creation week; and that the days of the creation week were simply twenty-four hours in length. Therefore, being land dwellers, dinosaurs were created along with man. These Biblical facts we have established.

NOTE TO THE READER: *It will be important for the reader to note the relationship between divine covenant and the length of a "day" in Jeremiah 33:19-21. First, the passage specifically tells us that God made a covenant between day and night to govern their length. I suggest this* *covenant was initiated in Genesis 1:5 when He named the light day and the darkness night.*

The strength of the covenant is highlighted in Jeremiah by Jehovah God when He says, "If you can break My covenant with day and My covenant with night, so that there will not be day and night in their season (or at their "appointed time," NIV), then My covenant may also be broken with David My servant, so that he shall not have a son to reign on his throne..." This says if the length of a day or a night as established by this covenant can be forced out of its covenantal bounds, the ultimate effect would be that Jesus Christ would not sit on David's throne (Remember, Scripture tells us that great David's greater Son will reign on the Earth forever, Isaiah 9:7, Daniel 2:44, Luke 1:32-33). Of course this covenant will not be broken, but strategic here is the fact that Christ's reign on Earth at His second coming is inseparably tied to this covenant concerning the length of day and night.

It is significant that Jesus asked His disciples in John 11:9: "Are there not 12 hours in the day?" This was rhetorical, having an obvious answer. Yes, there are 12 hours in the day and Jesus knew this because He made the original covenant. Thus, the first day of the creation week was 24 hours in length and all other solar days since have been 24 hours in length by divine covenant.

Job 40:15-24

I believe Job 40:15-24 identifies the "behemoth" to be an example of the great dinosaurian sauropods (the Hebrew "behemoth" being the plural form of the singular "behemah"). If this can be established, it is significant that the passage says they were "made along with man." However, since the Book of Job was authored about 300 years after the Flood and if the behemoths were sauropods, their presence demands that their progenitors were on the Ark of Noah.

The majority of the publisher's notes in many modern versions of the English Bible (for example the New International Version), tell us that the behemoth was most likely an elephant or a hippopotamus. It is telling that this intimation is made in the modern English translations and is not as pronounced in the older versions of the English Bible. I would suggest this is due to the fact that Darwinian thinking has thoroughly penetrated modern Biblical scholarship at an alarming rate. They add these questionable notes for a number of reasons. One of which is that it is much easier to explain the existence of elephants and hippos dwelling with man than dinosaurs. Dr. Henry Morris corroborates this fact:

> Modern Bible scholars, for the most part, have become so conditioned to think in terms of the long ages of evolutionary geology that it never occurs to them that mankind once lived in the same world with the great animals that are now only found as fossils. (Morris, 1988, p. 115)

Nevertheless, whatever behemoth happens to be—dinosaur, elephant, or hippo—it was "made with man" according to Job 40:15. This is the salient fact of this verse.

In verse 16, the writer tells us that behemoth had a unique anatomical advantage due to their specially designed "loins," as well as having exceptionally strong stomach muscles. The word "loins" is derived from the Hebrew "motnayim," meaning the "hips or the lower part of the back, i.e. the middle of the body" (Harris et al., 1980, 1, pp. 536-537). The consensus among scholars is that this word is more a reference to the small of the back or the unique tendon/ligament connection in this section of the lower back than to anything else.

D. Dixon, B. Cox, G. Savage and B. Gardiner tell us of a remarkable feature discovered in Brachiosaurian backbone anatomy, which very well may help to explain the unique flexibility and strength of the behemoth "loins," or the lower back. They explain:

> The secret of supporting such a massive body lay in the construction of Brachiosaurus' backbone. Great chunks of bone were hollowed out from the sides of each vertebra, to leave a structure, anchor-shaped in cross section, made of thin sheets and struts of bone. The resulting skeleton was a master-piece of engineering—the lightweight framework, made of immensely strong, yet flexible, vertebrae, each angled and articulated to provide maximum strength along lines of stress. (Dixon et al., 1988, pp. 128-129)

Interesting isn't it.

Verse 17 casts a great deal of light on the proper identification of this animal when it says it "moved or swayed its tail like a cedar" (NIV, "his tail sways like a cedar"). The Hebrew word "cedar" means to be tenacious at its roots and seems to be a reference to the cedars of Lebanon. These trees are exceedingly large, they are mentioned in the Old Testament, and, of course, are native to the Middle East. They can reach heights of 130 feet or higher with a trunk diameter of more than 8 feet. They are huge!

Thus, comparing the tail of this animal to a cedar of Lebanon means that it must have been vastly larger than an elephant's tail or a hippopotamus' tail—they are mere twigs by comparison. The tail being described in the text is more in line with a Brachiosaurus tail or some other kind of sauropod.

Moreover, verse 18 describes the bones of this animal as "beams of bronze" and "bars of iron" which explains, I think, the massive size of the bones of a sauropod. For example, Brachiosaurus fossil bones have been discovered to be enormous, with humerus bones (foreleg bones) to be 6 feet 4 inches in length and femur bones (hind leg bones) to be 6 feet 3 inches in length (Ham, 1998, p. 71).

In verse 19, behemoth is also called the "first of the **ways** of God" in the NKJV, and the NIV says the animal "ranks first among the **works**

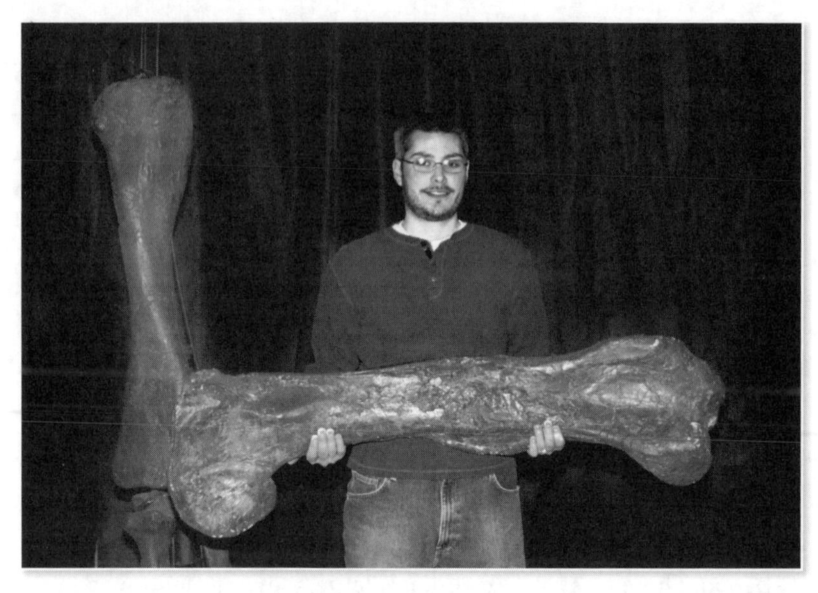

Edmontosaurus fossil leg bone that helps us understand Job 40:18

of God." This statement clearly shows that this animal was the greatest land animal God ever created; it was among the "first" or the "greatest" of His terrestrial creation. The only reasonable conclusion is that the animal described in this passage was dinosaurian—most likely a Brachiosaur, a Diplodocus, an Ultrasaurus, or maybe an Apatosaurus. It, quite obviously, could not have been an elephant or a hippo!

Since the book of Job is considered the oldest book of the Bible and was most likely written just before or during the life Abraham, it would have been written sometime about 250 to 300 years **after** the Flood. Thus, this passage reveals the presence of dinosaurian creatures after the Flood and is a Biblical confirmation of their presence in the Ark of Noah (Whitcomb, & Morris, 1961, pp. 23-35; also cited in Morris, 1988, pp. 12-16).

Consider this: In 1990, a research team from the University of Montana exhumed a nearly complete fossilized skeleton of a Tyrannosaurus rex. After the fossil had been unearthed, the team moved it to the university laboratory for continued preparation and inspection. Dr. Mary Schweitzer (at the time a post-graduate student) led the research. It was soon discovered that "parts deep inside the long bone of the leg had not completely fossilized"(Schweitzer, & Staeder, 1997, pp. 55-56).

Thinking that it may be possible to find some biomolecules, Schweitzer prepared a small fragment of this bony tissue for microscopic inspection. She reported that the:

> ...thin slice of a T. Rex bone glowed amber under the lens of my microscope. Blood vessel channels snaked through a bone matrix, and tiny chambers known as lacunae, which house bone-forming cells, appeared as small ovals.
>
> One by one my co-workers, paleontology students at Montana State University, took turns peering through the eyepiece. The lab filled with murmurs of amazement, **for I had focused on something inside the vessels that none of us had ever noticed before: tiny red objects, translucent red with a dark center.**
>
> **Then a colleague took one look at them and shouted, "You've got red blood cells. You've got red blood cells!"**
>
> Red blood cells? The shape and location suggested them, but blood cells are mostly water and **couldn't possibly have stayed preserved in the 65-million-year-old tyrannosaur.** Perhaps the mysterious structures were, at best, derived from blood, modified over the millennia by geological processes....
>
> So I showed these microscopic bones slices to my boss, a paleontologist Jack Horner, renowned for his work on dinosaur nesting sites. He took a long look and then asked, "So you think these are red blood cells?" I said, "No." He said, "Well, prove that they're not."

> [She further reported that after six to seven years of research.] **So far we haven't been able to.** (Schweitzer, & Staeder, 1997, p. 55; emphasis added)

This evidence creates two enormous problems for the so-called "millions of years" dinosaur age. By Dr. Schweitzer's own admission, "blood cells...couldn't possibly have stayed preserved in the 65-million-year-old tyrannosaur...." That is, if the T. rex bone is really 65 million years old.

Professor Jack Horner, Schweitzer's boss, wrote a book of his own in 1997 entitled, *Dinosaur Lives.* On page 228 he plainly admits:

> In another effort to make fossils speak in new ways, post-graduate Mary Schweitzer has been trying to extract DNA from the bones of T. rex. Originally...she had intended to thin-section the bones and conduct a histologic [tissue] investigation. **But under the microscopes there appeared to be blood cells preserved within the bone tissue.** Mary conducted a number of tests in an attempt to rule out the possibility that what she'd discovered were in fact blood cells. **The tests instead confirmed her initial interpretation.** (Horner, & Dobb, 1997, p. 228; emphasis added).

How could organic tissue last 65 million years and not completely decay? It couldn't, that is if it is really 65 million years old! Since this 1997 report, there has been much research concerning this subject. To illustrate the extent of this issue, I want to paraphrase some of the dialogue from the script of a July 24, 2007, television broadcast of NOVA Science NOW.

Peter Standring, a Nova correspondent, introduced Dr. Mary Schweitzer as a Paleobiologist with North Carolina State University by telling the viewing audience that Dr. Schweitzer "defined" her career while she was preparing a large Tyrannosaurus rex specimen when she was still in Montana. Explaining this reference, Schweitzer indicated that while she was working on a fossil leg bone of this T. rex, she saw "a bunch of stuff." This "stuff" was actually matter that she observed in the microscopic slide that was initially identified as "vessels and red blood cells."

Dr. Schweitzer then explained that, "Inside those channels where the blood vessels would run were these little round red structures that were all kind of lined up like a train. And they were bright and red and translucent." She said, "Nobody else had seen anything like that before."

Later, Jack Horner, her mentor at Montana State University, unearthed another T. rex and she again found soft tissue anatomy. In fact, she has repeated this procedure in many other dinosaur fossils and has found soft, organic tissue. Dr. Schweitzer said she found "Blood vessels, transparent, hollow, pliable, flexible branching blood vessels that contain small, round, red microstructures floating in the vessels."

She said, "This is not possible. Do it again." We got another piece of bone, we put it in the solution, we waited two or three or four weeks, looked again...more blood vessels. We must have repeated that with probably 17 or 18 different fragments of bone" (Fine, 2007, NOVA/ Transcripts/NOVA science NOW).

The question remains: Can soft tissue anatomies survive for millions of years, regardless of the preservation process? There is nothing in science that explains this dilemma. For sure, it is not something that we have ever seen before, that is, if the specimens are really millions of years in age.

However, this is not the first confirmed report of non-fossilized material associated with dinosaurs. Fresh dinosaur bones, not fossil bones, have also been found.

Dr. Margaret Helder, Science Editor for the *Reformed Perspective* magazine, indicated in 1961 a petroleum geologist found a bone bed in Northwest Alaska that was about two-feet deep. Thinking the fresh bone material to be recently processed buffalo bones, he gathered some of them but essentially ignored them. Some 20 years later these bones were properly identified as duckbill, horned, and carnivorous dinosaur bones. Dr. Helder said that scientists from the University of Alaska and the University of California at Berkeley have begun quarrying in this bone bed (Helder, 1992, pp. 16-17; also cited by Kyle L. Davies, 1987, pp. 198-200).

Did you hear that?—fresh bones, non-fossilized bones of all sorts of dinosaurs. Maybe they are not as old as they have been advertised.

This is remarkable evidence for the Biblical model of the recent creation of man and all land animals (including dinosaurs) on the same day. It also supports the reality of a recent worldwide Flood that caused powerful forces and process rates exceeding anything known to man. Hence, the Bible's position on man and dinosaurs is clear; they were created on the same day and those placed on the Ark survived the Flood. Some animals have become extinct, some have not. There is much evidence that confirms these Biblical assumptions found all over the world. So...

What Does All of This Mean?

Uniquely for the Bible believer, **it means** that an aggressive, yet egregious, evolutionary story has been sold to the American public for the last 100 years. For the last 49 years, the evolutionary crowd has totally controlled the public school offering by not permitting any form of organized curriculum delivery to the student body that presents the arguments for design or Biblical creation. This is a viral form of censorship and dogmatic indoctrination with a vengeance.

The evolutionary story about dinosaurs is just another chapter in the state-adopted religion of evolution that rejects God and the Bible. Even though many scientists espouse evolution, remember, there is a vast difference between a scientific finding and the opinion of a scientist. The evolutionary story about dinosaurs is mere opinion—nothing more!

It means that the Genesis account of history is remarkably accurate, and that the predictions based on the Genesis Model are in greater agreement with what we observe in the real world than are the predictions of evolution.

It means that the earth is trying to inform us that its rock layers, fossils, caves, canyons, uplifted folds and dozens of other features are the result of a global catastrophic event; and the only such event known in Biblical history is the Genesis Flood!

It means that the missing dinosaurs are an open testimony to the sudden disappearance of millions of life forms—all disappearing at the same time—all from their own habitat! Thus, this gives remarkable testimony to an incredible catastrophic event in the recent past.

More importantly, **it means** that every time we check what Jesus said about the earth, it perfectly checks and corresponds with the assumptions of creation and global catastrophe. **And this means** that what Jesus said about heaven is also true!

It means that all of the present evidence pointing to a worldwide Flood also declares the immediate feasibility of another worldwide event. This time a judgment of fire!

Since the people of Noah's day could only have been saved by boarding the Ark, the only way to be saved from this next event is to be in the Ark. And since the **Ark** today is not a boat, but a **Person,** the Lord Jesus Christ—**it means** that salvation can only be secured by being in that **Ark.** The great Creator God became our Savior and is now our **ARK!**

For as it was in the days of Noah, so shall it be in the day of the coming of the Son of Man. (Matthew 24:37)

The Biblical view of dinosaurs can only be properly understood in light of the Creator and His plan for man, as recorded in the Bible!

References

Achenbach, J. (2004, March). Dinosaurs Come Alive. *National Geographic,* 203(3).

Ager, D. (1976). The Nature of the Fossil Record. *Proceedings of the British Geological Association,* 87(2).

Ager, D. (1993). *The New Catastrophism.* Cambridge, UK: Cambridge University Press.

Austin, S. A. (1994). *Grand Canyon: Monument to Catastrophe.* Santee, CA: Institute for Creation Research.

Austin, S. A. (2000, March). Archaeoratpor: Feathered Dinosaur from National Geographic doesn't Fly. *Impact.* Dallas, TX: Institute for Creation Research.

Baker, H. B. (1938). Uniformitarianism and Inductive Logic. *Pan-American Geologist,* 69.

Barnhart, R. K. (1986). *Dictionary of Science.* Maplewood, NJ: Hammond, Inc.

Barzun, J. (1958). *Darwin, Marx, Wagner.* Garden City, NY: Doubleday.

Behrensmeyer, A. (1984, November/December). Taphonomy and the Fossil Record. *Scientific American,* 72.

Booher, H. R. (1998). *Origins, Icons and Illusions.* St. Louis, MO: Warren H. Green, Inc.

Bowden, M. (1982). *The Rise of the Evolution Fraud.* San Diego, CA: Creation-Life Publishers.

Bozarth, G. R. (1978, February). The Meaning of Evolution. *American Atheist,* 20(2).

Broadhurst, F. M. (1964, Summer). Some Aspects of the Paleontology of Non-Marine Faunas and Rates of Sedimentation. *American Journal of Science,* 262.

Coffin, H. (1983). *Origin by Design.* Hagerstown, MD: Review and Herald Publishing.

Colbert, E. (1965). *Age of Reptiles.* New York: Norton.

Colbert, E. (1968). *The Great Dinosaur Hunters.* New York: Dover Publications.

Colbert, E. (1968). *Men and Dinosaurs.* New York: E.P. Dutton and Co.

Colbert, E. (1983). *Dinosaurs: An Illustrated History.* Maplewood, NJ: Hammond Inc.

Darwin, C. (1998, Sixth Ed.). *The Origin of Species.* New York: The Modern Library.

Darwin, F., & Seward, A. C. (Eds.). (1972). *More Letters of Charles Darwin* (Vol. 2). London: John Murray.

Davies, K. L. (1987, January). Duckbill Dinosaurs (Hadrosauridae and Ornithisschia) from the North Slope of Alaska, *Journal of Paleontology,* 61(1).

Davis. D. D. (1949). Comparative Anatomy and the Evolution of Vertebrates. *Genetics, Paleontology and Evolution.* Princeton University Press. Retrieved July 2, 2009, from http://www.pathlights.com/ce_encyclopedia/sci-ev/sci_vs_ev_12b.htm

Dinosaurs. (1987). World Book Publication.

Dixon, D., Cox, B., Savage, R. J. G., & Gardiner, B. (1988). *The MacMillan Illustrated Encyclopedia of Dinosaurs and Prehistoric Animals: A Visual Who's Who of Prehistoric Life.* New York: Macmillan Publishing Co.

Doolan, R. (1993, September-November). Are Dinosaurs Alive Today? *Creation Ex Nihilo,* 15(4).

Dunbar, C. O. (1960, Second Ed.). *Historical Geology.* New York: John Wiley & Sons.

Engel, A. (1969). Time and the Earth. *American Scientist,* 57.

Fedducia, A. (1993, February 5). Evidence from Claw Geometry Indicating Arboreal Habits of Archaeopteryx. *Science,* 259.

Fine, S. (Executive Producer). (July 24, 2007). *NOVA scienceNOW* [Television broadcast]. WGBH/Boston by NOVA: WGBH Educational Foundation. Retrieved July 2, 2009, from http://www.pbs.org/wgbh/nova/transcripts/3411_sciencen.html

Futuyma, D. J. (1982). *Science on Trial.* New York: Pantheon Books.

George, T. N. (1960, January). Fossils in Evolutionary Perspective. *Science Progress,* 48.

Gish, D. (1992). *Dinosaurs by Design.* Green Forest, AR: Master Books.

Gish, D. (1995). *Evolution: The Fossils Still Say No!* El Cajon, CA: Institute for Creation Research.

Gould, S. J. (1977, May). Evolution's Erratic Pace. *Natural History,* 86.

Gould, S. (1977). The Return of Hopeful Monsters. *Natural History,* 86.

Gould, S. (1986, January). A Short Way to Bug Ends. *Natural History,* 95.

Gretener, P. E. (1967, November) Significance of the Rare Event in Geology, *Bulletin,* 51, American Association of Petroleum Geologists.

Ham, K. (1998). *The Great Dinosaur Mystery Solved.* Green Forest, AR: Master Books.

Harris, R. L., Archer, G. L., & Waltke, B. K. (Eds.). (1980). *Theological Wordbook of the*

Old Testament (Vol. 1). Chicago: Moody Publishers.

Harris, R. L., Archer, G. L., & Waltke, B. K. (Eds.). (1980). *Theological Wordbook of the Old Testament* (Vol. 2). Chicago: Moody Publishers.

Helder, M. (1992, June/July). Fresh Dinosaur Bones Found. *Creation Ex Nihilo,* 14(3).

Heylmun, E. B. (1971, January). Should We Teach Uniformitarianism? *Journal of Geological Education,* 19(1).

Himmelfarb, G. (1996). *Darwin and the Darwinian Revolution.* Chicago: I.R. Dee.

Hitching, F. (1982). *The Neck of the Giraffe.* New Haven: Ticknor & Fields.

Horner, J. R., & Dobb, E. (1997). *Dinosaur Lives: Unearthing an Evolutionary Saga.* New York: Harper/Collins Publishers.

Hsu, K. J. & McKenzie, J. A. (1986, March). Rare Events in Geology Discussed at Meeting. *Geotimes,* 31.

Lampton, C. (1989). *New Theories on the Dinosaurs.* New York: Franklin Watts.

Lewin, R. (1988, July 15). A Lopsided Look at Evolution. *Science,* 241.

Lewontin, R. (1981, October 22). The Inferiority Complexity. [Review of the book The mismeasure of man]. *New York Review of Books,* 28(16). Retrieved July 30, 2009, from http://www.nybooks.com/articles/article-preview?article_id=6867

Lewontin, R. (1997, January 9). Billions and Billions of Demons. [Review of the book The demon-haunted world: Science as a candle in the dark]. *The New York Times Book Reviews.* Retrieved July 30, 2009, from http://www.drjbloom.com/Public%20 files/Lewontin_Review.htm

Lyell, K. (1881). *Life, Letters and Journal of Sir Charles Lyell,* (2). London: John Murray.

Mackal, R. P. (1987). *A Living Dinosaur: In search of Mokele-Mbembe.* New York: E. J. Brill.

Matthews, L. H. (1972). Introduction in *Origin of Species.* London: J. M. Dent & Sons.

Marton, A. (1985). What is Uniformitarianism, and How Did It Get Here? *Horus 1*(2).

Morris, H. M., & Whitcomb, J. C. (1961). *The Genesis flood.* Philadelphia: Presbyterian & Reformed Publishing Co.

Morris, H. M. (1984). *The Biblical Basis for Modern Science.* Grand Rapids, MI: Baker Book House.

Morris, H. M., & Parker, G. (1987). *What is Creation Science?* Green Forrest, AR: Master Books.

Morris, H. M. (1988). *The Remarkable Record of Job.* Green Forest, AR: Master Books.

Morris, H. M. (1989) *The Long War Against God.* Grand Rapids, MI: Baker Book House.

Morris, H. M., & Morris, J. (1996). *The Modern Creation Trilogy* (Vol. 2). Green Forest, AR: Master Books.

Morris, H. M. (1997). *That Their Words May Be Used Against Them.* Green Forest, AR: Master Books.

Morris, H. M. (2000). *The Long War Against God.* Green Forest, AR: Master Books.

Morris, H. M. (2000, November 1). *The Profusion of Living Fossils.* Retrieved June 30, 2009, from http://www.icr.org/article/profusion-living-fossils/

Morris, J. (1994). *The Young Earth.* Green Forest, AR: Master Books.

Nelson, G. (1971). Origin and Diversification of Teleostean (bony) Fishes. *Annals of the New York Academy of Sciences.*

North, G. (1994) The Crisis of the Old Order. Retrieved August 10, 2009, from http://www.garynorth.com/freebooks/docs/a_pdfs/newslet/position/9405.pdf

Olson, S. L. (1999, November 1). Open letter to Dr. Peter Raven, Secretary, National Geographic Society.

O'Rourke, J. E. (1976, January). Pragmatism Versus Materialism in Stratigraphy. *American Journal of Science,* 276.

Oxburgh, E. R. (1983, September 1). Sunset on the Great Dinosaur Disaster. *New Scientist,* 99(1373).

Park, P. (1993, January 23). Science: Beast from the Deep Puzzles Zoologist. *New Scientist,* 1857.

Parker, G. (2006). *The Fossil Book.* Green Forest, AR: Master Books.

Patterson, C. (1981). *The Listener,* 106. British Broadcasting Company.

Patterson, C. (1999). *Evolution* (2nd Ed.). London: Natural History Museum.

Prothero, D. (2004). *Bringing Fossils to Life.* Boston: McGraw Hill.

Sattler, H. R. (1983). *The Illustrated Dinosaur Dictionary.* New York: Lothrop, Lee & Shepard Books.

Schweitzer, M., & Staeder, T. (1997, June). The Real Jurassic Park: The Blood of a T. rex Won't Bring the Dinosaur Back to Life! *Earth,* 6(3).

Simpson, G. G. (1953). *The Major Features of Evolution.* New York: Columbia University Press.

Selznick, Blazer, McCormack, Newton, & Rasmussen. (1985 & 1988). *Biology.* Glenview, IL: Scott, Foresmen, & Co.

Sloan, C. P. (1999). Feathers for T-rex. *National Geographic,* 196.

Spamer, E. (1989). The Development of Geological Studies in the Grand Canyon. *Tryonia* 17.

Spieker, E. M. (1956, August). Mountain-building Chronology and Nature of Geologic Time Scale. *Bulletin of American Association of Petroleum Geologists.*

Sunderland, L. (1984). *Darwin's Enigma.* Green Forrest, AR: Master Books.

Perloff, J. (1999). *Tornado in a Junkyard.* Arlington, MA: Refuge Books.

Raup, D. (1979). *Field Museum of Natural History Bulletin,* 50.

Raup, D. (1981, July 17). Evolution and the Fossil Record. *Science,* 213.

Ridley, M. (1981, June 25). Who Doubts Evolution? *New Scientist,* 90.

Rigby, S. (1993, March 18) Graptolites Come to Life. *Nature,* 362.

Valentine, J. (1966, May). The Present is the Key to the Past. *Journal of Geological Education,* 14.

Valentine, J. W., & Erwin, D. H. (1987). Interpreting Great Developmental Experiments: The Fossil Record. *Development as an Evolutionary Process.* New York: Allen R. Liss.

Webster, D. (2000, June), Debut Sue: Chicago's Field Museum Unveils the World's Most Famous T. rex. *National Geographic,* 197(6).

Weishampel, D., Dodson, P., & Osmolska, H. (Eds.). (1990). *The Dinosauria.* Berkley, CA: University of California Press.

Weishampel, D., Dodson, P., & Osmolska, H. (Eds.). (2007). *The Dinosauria.* Berkley, CA: University of California Press.

West, R. R. (1968, May). Paleoecology and Uniformitarianism. *Compass,* 45.

Whipple, F. L. (1978). The Earth as Part of the Universe. *Annual Review of Earth and Planetary Sciences,* 6.

Whitcomb, J. C., & Morris, H. M. M. (1961). *The Genesis Flood: The Biblical Record and Its Scientific Implications.* New Jersey: P&R Publishing.

Woodmorappe, J. (1981, June). The Essential Nonexistence of the Evolutionary-Uniformitarian Geologic Column: A Quantitative Assessment. *Creation Research Society Quarterly,* 18(1).

Index of Notable People Mentioned

Index of Notable Terms

Index of Notable Groups and Organizations

SERIES OVERVIEW

Over two years in the making, The War of the Ages series features footage from around the globe of historic locations that played a significant role in the history of mankind. Visit ancient Greek ruins, the birthplace of Charles Darwin, battlefields from the Revolutionary War, and many other interesting places as we uncover the truth behind the battles in the ongoing War of the Ages.

Grade 3 Teacher's Edition
978-1-933267-54-8

Grade 4 Teacher's Edition
978-1-933267-70-8

Teacher's Editions are only $47.95

Grade 5 Teacher's Edition
978-1-933267-86-9

Grade 6 Teacher's Edition
978-1-60725-586-4

Get the Total Package for $98.80 includes Teacher's Edition, Student Text, Tests & Quizzes, Answer Key and Multimedia/Support DVD

Grade 3 978-1-933267-51-7

Grade 4 978-1-933267-67-8

Grade 5 978-1-933267-83-8

Grade 6 978-1-933267-99-9

EPISODE 1
What is Worldview?

EPISODE 2
The Foundation of Worldview

EPISODE 3
The Pursuit of Truth

EPISODE 4
The Impact of Evolutionism on American Christians

Creation Truth
FOUNDATION
TM